OIL AND WILDERNESS IN ALASKA

Selected Titles in the American Governance and Public Policy Series

Series Editors

GERARD W. BOYCHUK, KAREN MOSSBERGER, AND MARK C. ROM

Oil and Wilderness in Alaska

Natural Resources,
Environmental Protection,
and National Policy Dynamics

GEORGE J. BUSENBERG

GEORGETOWN UNIVERSITY PRESS
Washington, DC

TD
195
.P4
B89
2013

Library of Congress Cataloging-in-Publication Data

Busenberg, George J.
    Oil and wilderness in Alaska : natural resources, environmental protection, and national policy dynamics / George J. Busenberg.
        pages cm— (American governance and public policy series)
    Includes bibliographical references and index.
    ISBN 978-1-58901-662-0 (pbk. : alk. paper)
    1. Petroleum engineering—Environmental aspects—Alaska.   2. Endangered ecosystems—Alaska.   3. Energy policy—United States.   4. Petroleum industry and trade—Political aspects—United States.   I. Title.
TD195.P4B89   2013
333.8′23209798—dc23

                                                                    2012049966

This book is printed on acid-free paper meeting the requirements of the American National Standard for Permanence in Paper for Printed Library Materials.

15 14 13   9 8 7 6 5 4 3 2   First printing

Printed in the United States of America

CONTENTS

ILLUSTRATIONS

Figures

Maps

ACKNOWLEDGMENTS

THIS BOOK is the product of many years of research concerning oil and wilderness in Alaska. During those years my research has benefited greatly from the advice and assistance of a number of people whom I would like to thank here. The theory of the policy process applied in this book was developed by Frank Baumgartner and Bryan Jones, and I am grateful for the efforts of these two scholars both in developing a new theory of the policy process and in supporting my efforts to apply that theory to empirical cases in environmental policy. I am grateful to the scholars Thomas Birkland, Peter May, Frank Laird, Monty Hempel, the late Frances Lynn, and the late William Freudenburg for sharing various insights that contributed to my efforts in developing this research project over the years. I thank Melanie Lewis and Daniel Hattrup for providing research assistance with the initial stages of this project. I am grateful to the Georgetown University Press editors who assisted me in the development of this book, including Don Jacobs, Gerry Boychuk, Karen Mossberger, Mark Rom, Richard Brown, Barry Rabe, and Gail Grella. I also thank the anonymous reviewers of the book manuscript, who provided clear and highly constructive advice on the manuscript that helped me to improve the final book. I am grateful to Christopher Robinson, who created the maps for this book.

I am deeply grateful to Soka University of America for providing faculty research funding and other institutional support that proved essential in the completion of this book project. I thank all of the students at Soka University of America who worked as research assistants in the development of this project, including Monika Mann, Heidi Helgerson, Elizabeth Guthrey, Jennifer Callahan, Yona Yurwit, Sarah Randolph, and James Cole Gauthier. I also want to thank Gosha Domagala and Leigh Moynihan at the Daisaku and Kaneko Ikeda Library at Soka University of America for helping me access a massive collection of scholarly works and primary data in support of this research project through interlibrary loan services.

This book incorporates materials from two previously published refereed articles in the journal *Review of Policy Research* by permission from Wiley-Blackwell (Busenberg 2008, 2011). I thank Wiley-Blackwell, *Review of Policy Research,* and the Policy Studies Organization for allowing use of this previously published material. I also thank the anonymous reviewers of these two articles for providing useful advice that contributed first to the improvement of the articles and subsequently to the improvement of this book.

This study was supported in part by grants from the U.S. National Science Foundation (NSF SBR-9520194, NSF SBR-9710522). The views expressed in this book are the sole responsibility of the author.

ACRONYMS

ADEC	Alaska Department of Environmental Conservation
ANILCA	Alaska National Interest Lands Conservation Act
BLM	U.S. Bureau of Land Management
CIRCAC	Cook Inlet Regional Citizens Advisory Council
GPS	Global Positioning System
MARPOL 73/78	International Convention for the Prevention of Pollution from Ships
NOAA	U.S. National Oceanic and Atmospheric Administration
OPA 90	Oil Pollution Act of 1990
OPEC	Organization of Petroleum Exporting Countries
OSRI	Prince William Sound Oil Spill Recovery Institute
PWS	Prince William Sound
PWS RCAC	Prince William Sound Regional Citizens' Advisory Council
RCAC	Regional Citizens' Advisory Council
TAPS	Trans-Alaska Pipeline System
USDA	U.S. Department of Agriculture

Introduction

THIS BOOK EXAMINES the development of national policies for oil development and nature conservation in the state of Alaska. Alaska contains the largest developed oil field in the United States, and oil development poses major threats to the environment across large areas of Alaska (McBeath et al. 2008). Alaska also contains the most extensive system of land conservation areas of any state in America, and national efforts to conserve the extraordinary wilderness and wildlife values of Alaska have repeatedly come into conflict with plans for oil development in the state (Miles 2009; Naske and Slotnick 2011). The issues of oil development and environmental protection in Alaska have been the focus of longstanding conflicts between environmental interests seeking to protect the Alaskan environment and development interests seeking to exploit Alaska's vast natural resources. The conflicts between these competing interests have contributed to national policy reforms that have established enduring systems for both oil development and environmental protection in Alaska. The central purpose of this book is to provide a theoretically driven examination of three national policy reform efforts that authorized the development of the nation's largest oil field in northern Alaska, established a vast system of protected natural areas in Alaska, and reformed the environmental management of the marine oil trade in Alaska to reduce the risk of oil pollution in that trade. The enduring institutional legacies and policy consequences of each reform period are examined, with a focus on the consequences of reform efforts for environmental protection. This book also examines the national and international repercussions of these reforms and the continuing policy conflicts concerning oil development and nature conservation in Alaska left unresolved by these reforms.

This book is designed to add a theoretical perspective to the existing scholarly literature on Alaska—a literature that is extensive, highly useful, and primarily empirical. This work aims to provide a concise and theoretically integrated account of select issues of environmental and natural resource policy in Alaska that have been the subject of extensive national interest and national reform efforts. This book is not intended to provide a comprehensive history of Alaska, and it is not intended to examine all of the myriad intricacies of reform or the biographies of the individuals involved in reform efforts. Instead, this book focuses on a theoretically driven examination of the interactions between ideas, interests, and institutions that have driven and sustained selected policy reforms in Alaska that are of national significance (Swedlow 2011). The central benefit of this theoretical approach is to establish a systematic framework for understanding the essential elements of the processes and consequences of policy change.

This book applies the punctuated equilibrium theory of policy change to examine three major periods of reform that shaped modern Alaska (Baumgartner and Jones 2009). The punctuated equilibrium theory has been widely recognized in the academic literature as an important theory of the policy process in general, and of environmental policy processes in particular (Repetto 2006). An overview of the literature on the theory is found in True, Jones, and Baumgartner (2007). The punctuated equilibrium theory predicts that the policy process includes distinct critical periods (punctuations) and equilibrium periods. In the context of the theory, a critical period is defined as a period of fundamental policy reform that establishes long-enduring institutional arrangements (such as new laws or new organizations) that cause a lasting realignment of actions in a policy domain. By contrast, an equilibrium period is characterized by incremental policy shifts. The term punctuated equilibrium therefore refers to a policy process containing both equilibrium periods, characterized by incremental policy adjustments, and punctuations, characterized by fundamental policy reforms with enduring consequences (Baumgartner and Jones 2009).

The pattern of policy change predicted by the punctuated equilibrium theory reflects the pivotal role of attention in the policy process. For example, reform at the national level typically depends on attention from the highest levels of national government. The collection of issues receiving attention at the highest levels of national government in a given time period forms the national agenda during that period. But opportunities to place a reform proposal on the national agenda are scarce, because government leaders can only attend to a few of the many policy issues under their authority at any given time. Proposals for reform in any given policy domain can therefore be crowded out of the national agenda for long periods of time; during these equilibrium periods, existing institutional arrangements maintain a stable framework of organization and understanding that tends to produce incremental policy change in the policy domain (Baumgartner and Jones 2009). In addition to the challenge of placing reform proposals on the national agenda, proponents of reform also frequently face the challenge of overcoming opposition to reform arising from groups that perceive threats to their interests and viewpoints in the proposed reforms. For example, opposition to proposed environmental reforms often arises from industries anticipating that those proposed reforms would increase industry costs due to environmental regulation or diminish industry access to natural resources. Conversely, environmental groups often oppose proposed reforms that would permit industrial activities that could cause environmental damage. In essence, groups that perceive an advantage in the existing state of affairs in a policy domain will typically oppose proposed reforms that would realign that state of affairs to their disadvantage (Layzer 2011).

Due to scarcity of attention on the national agenda and frequent abundance of opposition to reform, proponents of reform are often at a disadvantage in the policy process when compared to supporters of the existing state of affairs in a policy domain. Therefore, equilibrium periods are common and punctuations are sporadic in the policy process (Baumgartner and Jones 2009). However, shifts in public and political attention can create occasional opportunities for proponents of reform to place their reform proposals on the national agenda and overcome opposition to reform. These critical periods also create opportunities for proponents of reform to secure the establishment of new institutional arrangements that fundamentally realign a policy domain. New institutions established in a punctuation often prove durable in the subsequent equilibrium period even in the face of sustained opposition, as opponents of an established reform typically face the difficulties of achieving another punctuation if they hope to revoke or significantly diminish the reform they oppose. It is generally difficult to establish a major reform, but it is also generally difficult to reverse or erode a major reform. Punctuations place the opponents of reform at a disadvantage in the policy process, and punctuations therefore tend to establish long-lasting equilibria structured by the institutional legacies of punctuations. The pattern of punctuated equilibria that emerges from this dynamic interplay of social forces is marked by equilibrium periods in which barriers to major reform tend to protect the existing state of affairs in a policy domain; these barriers can occasionally be overcome by reformers who establish new institutions and enduring realignments of policy during critical periods of policy reform.

The punctuated equilibrium theory holds that new institutional arrangements established in critical periods are associated with supporting ideas referred to as policy images. A policy image is the collection of ideas that forms the conceptual basis for institutional reform. Policy images are significant because they inform the design and purposes of new institutional arrangements established in punctuations. A punctuation that establishes new institutions typically begins as a period during which alternative proposals for new institutional arrangements and associated policy images are recognized as important items on the government agenda and debated; the punctuation ends when one set of new institutional arrangements and a supporting policy image is formally established after prevailing over other alternatives considered in the debate. Because policy images provide the conceptual basis of reform, the formation of a policy image normally precedes the formation of institutional arrangements during punctuations. Proposed policy images can therefore drive institutional reform, and established policy images can exert significant influence over the actions of their associated institutions. In sum, the punctuated equilibrium theory predicts that critical periods of reform can establish new institutional arrangements and associated policy images in a given policy domain. Furthermore, the theory predicts that the new institutions and policy images established in critical periods can endure as institutional legacies in subsequent equilibrium periods, with enduring policy consequences (Baumgartner and Jones 2009).

The research question driving this book is whether the three key periods of policy reform examined here match the pattern of policy change predicted by the punctuated equilibrium theory. The first reform authorized the establishment of large-scale oil transportation infrastructures in Alaska. The second reform created a vast system of protected natural areas in Alaska. The third reform fundamentally realigned the systems used to protect the environment of Alaska from the hazard of marine oil pollution. This study hypothesizes that each of these three key reforms was accomplished through a critical period characterized by a rise in political and media attention in support of the reforms in question (the independent variables driving punctuations in the theory) and the establishment of enduring ideas and institutions that allowed the implementation of those reforms (the dependent variable in the theory). The evidence gathered in this study supports this hypothesis by identifying a pattern of policy change with equilibrium periods characterized by incremental policy adjustments and three punctuations characterized by a rise in congressional and media attention in support of reforms and the establishment of enduring new institutions and associated policy images that allowed those reforms to be implemented and sustained.

The reform efforts examined in this book were not confined to the state of Alaska, but rather were driven primarily by policy dynamics at the national level. The scope of research in this book therefore has a central focus on policy dynamics at the national level, while also considering relevant policy dynamics at the regional, state, and international levels. Research for this book was conducted over a period of eighteen years and included three research trips to Alaska, supported in part by funding from the National Science Foundation. The principal sources for this book include an extensive set of primary and secondary documents containing a highly detailed description of the events examined. Documents used in this study include the *U.S. Congressional Record*, congressional reports, presidential documents, government agency documents, *New York Times* articles, nongovernmental organization documents, industry documents, and the academic literature. The information in these documents was supplemented when needed through in-person, telephone, and e-mail interviews conducted by the author with seventy-eight respondents representing various organizations involved in Alaska policy issues (including government agencies, oil corporations, and nongovernmental organizations). An overview of the argument of the book is provided next.

An Overview of the Book

Three key policy reforms are examined in three separate chapters of this book. Chapter 2 examines a critical period of policy reform that allowed the development of the Trans-Alaska Pipeline System and also set the stage for other key policy reforms examined in subsequent chapters. The Trans-Alaska Pipeline System transports oil produced in the North Slope region of Alaska through an eight-hundred-mile pipeline and a marine oil terminal that loads oil onto tankers for transportation by sea to other regions (Coates 1993; McBeath et al. 2008). The transportation of large volumes of North Slope oil through the Alaska pipeline has created an enduring risk of environmentally damaging oil spills in Alaska. The critical period of reform that authorized the Alaska pipeline was preceded by an equilibrium period in this policy domain (beginning with Alaska statehood in 1959 and ending with the 1968 announcement of a large oil field discovery in northern Alaska) characterized by incremental policy changes concerning oil in Alaska, as well as limited congressional and media attention to the subject of oil in Alaska. This equilibrium period was also characterized by incremental policy changes concerning two subjects that would subsequently become the focus of major reform efforts—nature conservation in Alaska and the land claims of the indigenous peoples of Alaska. During this equilibrium period oil development occurred on a limited scale in Alaska, the indigenous peoples of Alaska made little progress in their land claims in Alaska, and federal conservation efforts in Alaska were not commensurate with the extraordinary opportunities for nature conservation then available in Alaska (an issue of national importance because Alaska was the largest state in America and contained the largest areas of wild land left in the nation). All three of these issues would receive far greater national attention and action in the subsequent Alaska pipeline punctuation that began with the announcement of a major oil field discovery on the North Slope of Alaska in 1968 and ended with the authorization of the Trans-Alaska Pipeline System project in 1973, following a protracted national policy conflict over the proposal for this pipeline project (Coates 1993; McBeath et al. 2008; Naske and Slotnick 2011; Ross 2000).

During the Alaska pipeline punctuation the prospects of Alaska's economy were greatly expanded by the discovery of the North Slope oil fields, but those expanded prospects depended on the development of a transportation infrastructure to move North Slope oil to the market. The federal government, the state of Alaska, and the oil industry pressed for the building of the Alaska pipeline to allow development of the North Slope oil fields. However, the Alaska pipeline proposal was contested by environmental groups concerned about the project's environmental impacts. Furthermore, the land claims of Alaska Native groups intersected the pipeline route and created a means for those groups to interfere with the pipeline project. The Alaska pipeline proposal thereby gave the indigenous peoples of Alaska an opportunity to press for the settlement of their land claims, and that settlement process also gave the environmental movement an opportunity to press for reforms that would create a systematic process of large-scale land conservation planning in Alaska. The new institutional arrangements created in this critical period established a major policy realignment that allowed significant advances for development, indigenous, and environmental interests in Alaska. This policy realignment was established by the approval of two federal laws: a 1971 law for Alaska Native land claims cleared the Alaska pipeline route, and a 1973 law authorized the pipeline project. The policy image supporting these new institutional arrangements reflected a multifaceted effort to promote oil development in northern Alaska, settle the Alaska Native land claims, and establish a systematic program to temporarily reserve vast areas of Alaska and allow the consideration of federal conservation options in those areas. The subsequent equilibrium period (1974 to the present) has largely perpetuated this institutional realignment and policy image both through large-scale oil development made possible by the construction and operation of the Trans-Alaska Pipeline System, and through the transfer of title of large areas of Alaska to the indigenous peoples of Alaska. The element of this realignment supporting a systematic process of conservation planning in Alaska was designed as a temporary measure; this measure served its intended function by setting the stage for subsequent reforms that established federal conservation areas across vast scales in Alaska, as described in chapter 3 (Coates 1993; McBeath and Morehouse 1994; Mitchell 2001; Ross 2000).

Chapter 3 examines a critical period of policy reform that authorized the creation of a very large system of protected natural areas in Alaska. This punctuation was preceded by an equilibrium period in the policy domain of Alaska lands conservation (1959–69) in which the scale of the federal nature conservation system in Alaska was overshadowed by the scale of the vast wild areas of Alaska that remained unprotected from development (Federal Field Committee for Development Planning in Alaska 1968). The beginning of the Alaska lands conservation punctuation in 1970 overlapped with the Alaska pipeline punctuation. The 1971 Alaska Native land claims legislation authorized a process of federal land conservation planning in Alaska that led to a prolonged policy conflict between environmental and development interests. During this period a national coalition of environmentalists sought the preservation of natural areas on a grand scale in Alaska, while development interests and the Alaska congressional delegation sought to maintain opportunities for economic development in many of the areas of Alaska that environmentalists sought to protect from development. This second critical period ended in 1980 with legislation that added approximately 105 million acres to federal conservation systems in Alaska, constituting the single greatest act of land conservation in the history of the United States and the world to that time (Allin 2008). The institutional arrangements structuring this conservation system included a mosaic of conservation units in Alaska that allowed a complex mix of land uses. The policy image supporting these new institutional arrangements reflected a multifaceted effort to conserve large areas of land in Alaska in a manner that protected both wilderness and wildlife, to allow the continuation of the longstanding practice of subsistence activities (such as hunting, fishing, and gathering) by the Alaska Natives and other residents of

Alaska within some conservation areas in Alaska, to permit mechanized access to Alaska conservation areas for traditional uses in recognition of the transportation difficulties imposed by the vast scale and limited road network of the state, to allow the continuation of sport hunting in some Alaska conservation units, and to create continuing opportunities for natural resource extraction in select areas of Alaska (Catton 1997; Nelson 2004). The subsequent equilibrium (1981 to the present) has perpetuated this conservation system through sustained management efforts that have a particular focus on balancing the protection of natural areas in Alaska with the perpetuation of recreation and subsistence activities in those areas (Miles 2009).

Chapter 4 examines a critical period of policy reform that fundamentally realigned the policies used to protect the environment of Alaska from the marine oil transportation system of the state. This critical period was triggered by the 1989 *Exxon Valdez* oil spill in Alaska, the largest marine oil spill in the history of the United States to that time (Burger 1997). The tanker vessel system for transporting oil from the Trans-Alaska pipeline commenced operations in 1977, in the midst of a national policy equilibrium regarding marine oil pollution marked by limited environmental regulation of the oil tanker trade in U.S. waters. During this equilibrium, large volumes of oil were transported through the coastal waters of Alaska using an oil tanker system with few safeguards against marine oil pollution. The *Exxon Valdez* disaster revealed the fundamental inadequacies of the safeguards against marine oil pollution in Alaska, and led to a sweeping reform of national marine oil pollution law in 1990 (Busenberg 2008; Clarke 1999; Grumbles and Manley 1995).

The *Exxon Valdez* oil spill punctuation established stronger national regulations of the marine oil trade in the maritime jurisdiction of the United States and created a series of new organizations designed to improve the environmental management of the marine oil trade in Alaska (Birkland 1997; Busenberg 2008; Randle 2012). The policy image supporting these new institutional arrangements reflected a multifaceted effort to mandate new design standards for oil tankers to reduce the risk of oil tanker spills nationwide, to increase the liability limits for oil spills to encourage safe practices in marine oil transportation nationwide, to create a new set of marine oil pollution standards and practices in Alaska to reduce the risks of the marine oil trade in Alaska, and to establish a system of progressive improvement in the safeguards against marine oil pollution in Alaska. The subsequent equilibrium period in this policy domain (1991 to the present) has perpetuated these institutional arrangements for environmental safety in the marine oil trade and has been marked by a sustained series of incremental enhancements to safeguards against marine oil pollution in Alaska (Busenberg 2008).

Chapter 5 examines the continuing conflict over oil development and environmental protection in Alaska in the wake of the three punctuations examined in the previous chapters. Chapter 5 focuses on national conflicts over oil development and environmental protection in the Arctic National Wildlife Refuge of northern Alaska and in the Arctic Ocean regions offshore of northern Alaska. Taken together, these cases demonstrate the continuing national importance of the conflicting pressures for natural resource development and environmental protection in Alaska that have shaped all of the policy reforms examined in this book. The concluding chapter summarizes the findings of this book and explores the wider theoretical implications of those findings.

In addition to examining the consequences of the three critical periods of reform for environmental protection in Alaska, this book also examines the national and international ramifications of these three reform periods. The Alaska pipeline punctuation allowed the development of a new and nationally significant domestic supply of oil for the United States, but also prompted unintended international repercussions. During the Alaska pipeline punctuation, the oil industry sponsored test voyages of an oil tanker in the Arctic waters off the coastlines of Alaska and Canada to assess whether these waters offered a route to ship oil from the North Slope of Alaska. Canada reacted to these Arctic tanker tests by claiming jurisdiction over pollution management standards in a large maritime area along the northern coastline of Canada despite strong U.S. resistance to such a large-scale expansion of Canadian jurisdiction in the Arctic Ocean (Elliot-Meisel 2009; Kirkey 1996). The Alaska pipeline punctuation marked the beginning of a series of reforms by which Canada progressively claimed environmental and economic control across a vast maritime region along the northern coastline of Canada (Grant 2010; Griffiths 1987). The Alaska lands conservation punctuation marked a key juncture in national conservation policies by vastly expanding the total areas of the national park, wildlife refuge, and wilderness preservation systems of America. Some of the conservation areas established or expanded in Alaska by this punctuation would connect with conservation areas in Canada to form large transnational protected areas. The Alaska lands conservation punctuation also held international ramifications by conserving large natural areas in Alaska that provided habitat for migratory species protected by international treaties between the United States and other nations (Docherty 2001). The *Exxon Valdez* oil spill punctuation held both national and international ramifications through the development of new standards for protective hull designs on oil tankers that reduced the risk of oil tanker spills. This critical period established a federal mandate for the progressive introduction of oil tankers with protective hull designs in U.S. maritime jurisdiction, setting the stage for a subsequent international effort that made protective hull designs the standard for oil tankers worldwide (National Research Council 1998, 26–30). In sum, each major period of reform examined in this book had policy implications that extended far beyond the borders of Alaska to encompass issues of national and international significance.

This study finds that each of the three critical periods of policy reform examined in this book match the patterns of reform predicted by the punctuated equilibrium theory. A sharp rise in congressional and media attention supportive of reform occurred during each of the three critical periods, and each critical period established enduring institutional arrangements and policy images with enduring policy consequences. As will be shown in the chapters to follow, the punctuated equilibrium theory provides a consistently accurate framework for understanding the dynamics of the reforms examined in this book. The first of these reforms is examined in the next chapter, which focuses on the development of the Trans-Alaska Pipeline System.

The Trans-Alaska Pipeline System

T HE CRITICAL PERIOD OF REFORM that eventually authorized the Trans-Alaska pipeline began with the announcement of the discovery of the Prudhoe Bay oil field in the North Slope region of Alaska in 1968. The Prudhoe Bay oil field was the largest oil field discovered in North America. In 1969, another oil field that ranked among the largest in North America (Kuparuk) was also discovered on the North Slope of Alaska (McBeath et al. 2008; Worster 1994). While a significant oil field discovery had occurred in the Cook Inlet region of Alaska in 1957, the North Slope oil fields contained much larger reserves of oil than Cook Inlet (McBeath and Morehouse 1994; McBeath et al. 2008). However, the remote North Slope region of Alaska was far less accessible for oil transportation than was Cook Inlet. The transportation of North Slope oil to the market would therefore be a major challenge for the oil industry.

The options for oil transportation from the North Slope included the use of oil tankers to ship oil by sea, the use of an oil pipeline, or both. A Trans-Alaska pipeline for North Slope oil was proposed in 1969 by a newly formed oil industry consortium called the Trans-Alaska Pipeline System (TAPS), which included three major oil corporations that together owned most of the oil discovered in the North Slope fields (Coates 1993; Coen 2012; McBeath et al. 2008). As an alternative to the Alaska pipeline proposal, the oil industry experimented with the idea of using oil tankers to transport oil directly from the North Slope region. But if tankers were to be loaded with oil at a marine terminal near the North Slope, those tankers would have to be capable of safely navigating through the ice often found in the Arctic waters along the northern coastline of North America.

The oil industry conducted tanker tests to assess the potential of an Arctic Ocean route for North Slope oil. These Arctic tanker tests consisted of two experimental voyages of an icebreaking oil tanker in Arctic Ocean routes. An icebreaking oil tanker with an armored hull was needed for such voyages because a conventional oil tanker could not safely sail in the Arctic sea ice (Coen 2012). The Arctic tanker tests were pursued by retrofitting the oil tanker *Manhattan* into an icebreaking oil tanker. The *Manhattan* had a heavily built structure and a powerful propulsion system that made it a good candidate for the Arctic tanker tests (Coen 2012, 34; Keating and Guravich 1970, 15). Due to the size of the ship and the project's tight schedule, the *Manhattan* was split into sections that were sent to different shipyards to be retrofitted with a layer of armor around the hull, internal structural reinforcements, a special icebreaking bow, and numerous other changes to prepare the tanker for its Arctic voyages (Coen 2012, 37–39; Keating and Guravich 1970, 89). The altered sections were then recombined to create a retrofitted *Manhattan* that was the largest icebreaker in history (Coen 2012, 80; Keating and Guravich 1970, 140).

The *Manhattan* was accompanied on its Arctic test voyages by American and Canadian icebreaking escort vessels (Coen 2012). During the first voyage the *Manhattan* successfully navigated the waters of the Arctic Ocean to and from the North Slope. The tests proved that an Arctic Ocean route for North Slope oil was feasible, but also indicated that the route would be costly and risky. A series of mishaps during the Arctic tanker tests demonstrated the inherent difficulties of the Arctic Ocean route. The *Manhattan* was repeatedly trapped in sea ice during the test voyages, and the *Manhattan's* hull was breached by a collision with ice. These mishaps indicated that even an icebreaking tanker could become trapped in sea ice or cause a major oil spill while navigating the northern coastline of North America. The difficult weather and ice conditions of the Arctic Ocean route could interrupt North Slope oil transportation, and the hull breach experienced by the *Manhattan* during the test voyages suggested a high risk of oil spills in this route as well. The Arctic tanker option would require the construction of a fleet of heavily armored icebreaking oil tankers and a fleet of icebreaking escort vessels to attempt the transport of large volumes of North Slope oil throughout the year. Furthermore, the loading of large oil tankers would require deeper waters than those found near the shores of the Prudhoe Bay oil field. The engineering options of dredging a deeper channel or building an offshore oil terminal in deeper waters posed severe technical difficulties in the Arctic conditions of the region. Marine oil shipping from the North Slope therefore posed major technical problems and environmental risks. In the end, the Arctic tanker tests suggested that an Arctic Ocean route for North Slope oil would be less reliable, costlier, and riskier than a pipeline system for North Slope oil (Coen 2012; Keating and Guravich 1970, 148).

Another proposal for the marine transportation of North Slope oil involved the use of submarine tankers, which could avoid the problems of icebreaking by navigating underneath sea ice along the route from the North Slope. This idea gained some credence from the experience of American submarines that had successfully navigated beneath Arctic sea ice (Elliot-Meisel 2009, 209; Grant 2010, 331–32). However, the notion of transporting North Slope oil using a fleet of large submarine oil tankers was untested. In addition to the novel challenges of engineering a new class of submarines to serve as large oil tankers, the submarine option also involved the challenge of building an Arctic marine oil terminal for submarines. Therefore, both of the Arctic shipping options for North Slope oil involved the challenge of building a marine oil terminal in very difficult Arctic circumstances (Coen 2012; U.S. Department of the Interior 1972).

In addition to the pipeline and marine route proposals, a number of other alternatives for the transportation of North Slope oil were proposed. Various proposals suggested moving the oil overland by trucks or trains, with the necessary supporting infrastructure of roads or rails made available to support further development in northern Alaska. Another proposal suggested that the oil could be moved by air using a fleet of very large airplanes. None of these alternatives was put to a full-scale test akin to the test voyages of the *Manhattan*, and indeed it appeared that some of the proposals (particularly those involving airplanes or trucks) were not feasible given the logistical difficulties posed by the anticipated scale of North Slope oil production. In essence, the alternative proposals for North Slope oil transportation were unproven at the scales and in the conditions found in northern Alaska. By contrast, the practice of transporting large volumes of oil using pipelines and large oil tankers had proven successful around the world (Coen 2012). The idea of building an overland oil pipeline for North Slope oil therefore prevailed over other alternatives for oil transportation considered during the critical period (Cicchetti 1972; Coates 1993; McBeath et al. 2008; Ross 2000; U.S. Department of the Interior 1972). The abandonment of the Arctic marine shipping options (and the particularly novel idea of moving oil by air) meant that the North Slope oil would have to travel overland across Alaska to reach markets outside of Alaska. However, overland oil transportation would carry sweeping policy implications for Alaska and the nation. The option that would eventually prevail was a lengthy oil pipeline that would cross a vast and contested landscape in Alaska. The policy reforms that eventually allowed the building of the Alaska pipeline would also cause a fundamental rearrangement of land management practices across Alaska (Busenberg 2011).

Two competing proposals emerged for the construction of an oil pipeline from the North Slope. The 1969 TAPS consortium proposal for a Trans-Alaska Pipeline System included a pipeline to transport oil from the North Slope to a marine oil terminal at an Alaskan seaport with fewer ice hazards than the North Slope. The oil would be loaded onto large oil tankers at the marine oil terminal for transportation to other regions. The proposed Trans-Alaska Pipeline System therefore included a land-based pipeline and a marine oil terminal, both of which would be under American jurisdiction (Coates 1993). A competing proposal included a much longer pipeline that would transport oil across both Alaska and Canada to the American market. This proposed Trans-Canada pipeline would eliminate the need for a marine oil transportation system, but would require building a very long pipeline that would be partly under Canadian jurisdiction (Cicchetti 1972; Coates 1993; Kirkey 1997).

The oil industry reached a swift consensus in favor of the Trans-Alaska Pipeline System over the many other alternatives for North Slope oil transportation. The Trans-Alaska Pipeline System appeared to offer good prospects for securing industry profits through a well-understood engineering design for maintaining a reliable flow of oil, an American route for the pipeline that minimized regulatory complexity and delay by avoiding Canada, and a marine transportation system that gave the oil industry the flexibility to transport North Slope oil to markets across the globe if the federal government approved (Cicchetti 1972; Coates 1993). However, the Trans-Alaska pipeline proposal included a pipeline route crossing federally owned lands in Alaska (McBeath et al. 2008). Federal actions were therefore needed to establish a route for the Alaska pipeline and to authorize construction of the pipeline. Those federal actions would be preceded by two major policy conflicts. One conflict centered on Alaska Native land claims that included lands in the route proposed for the Alaska pipeline. The other conflict centered on environmentalist objections to the Alaska pipeline project. The policy dynamics of both conflicts are examined in the sections that follow, beginning with an examination of the initial rush to prepare for the building of the Alaska pipeline.

The Rush to Prepare for the Alaska Pipeline

The early stages of the critical period examined in this chapter were character-ized by a rush to prepare for the Alaska pipeline by the oil industry and the governor of Alaska. What became evident in this initial pipeline rush was a lack of attention to the issues of environmental protection and appropriate engineering. A fundamental engineering challenge of the proposed Alaska pipeline was the presence of frozen soils (permafrost) along much of the pipeline's proposed route. The permafrost problem was clearly demonstrated during the early rush to prepare for the Alaska pipeline. Shortly after the discovery of the Prudhoe Bay oil field was announced, Alaska governor Wal-ter Hickel authorized the building of a road from the end of the existing public road system in Alaska north to the vicinity of Prudhoe Bay. This four-hundred-mile dirt road, built by the state of Alaska in the period 1968–70, was intended to forge an overland route to support the industrial develop-ment of northern Alaska. Essentially, the road was built by scraping off snow and vegetation along the surface of the route, which proved to be an ineffec-tive approach due to the seasonal thawing of exposed permafrost soils that turned the road into a ditch. The environmental damage caused by the road demonstrated the construction challenges posed by permafrost and became a rallying point for environmentalists opposing the Alaska pipeline project (Coates 1993, 164–67; Wayburn and Alsup 2004, 213).

The initial oil industry proposal for the Trans-Alaska pipeline called for a largely buried pipeline that was not well-suited for the permafrost soils along the route. North Slope oil would not only be hot when extracted from the oil fields but would also be heated by pressure and friction generated by oil flowing within the pipeline. The transfer of heat from the oil inside a pipeline buried in permafrost would thaw the surrounding frozen soils, thereby creat-ing the risk that melting soils would shift around the pipeline and cause the pipeline to break. Therefore, the buried pipeline design was fundamentally flawed from the perspective of both pipeline integrity and environmental safety (Coates 1993, 176–200, 256).

The rush by the oil industry to prepare for the building of the Alaska pipeline was evident in that the oil industry began initial road construction activities along a section of the intended route of the Alaska pipeline (and ordered the eight hundred miles of pipe needed for the project) in 1969 (Berry 1975, 104–5; Coates 1993, 178). The overall effect of these industry actions was to set the stage for the Alaska pipeline project in advance of federal approval for the project as a whole. These preparatory actions held the potential to both accelerate the project and create a sense of momentum that might sway public and political opinion in favor of federal approval for completion of the project. However, Alaska Native groups and environmental organizations interfered with these attempts to create momentum for the pipeline project. The next section examines the entry of Alaska Native groups and environmentalists into the Alaska pipeline debate.

Resolving the Alaska Native Land Claims

The early stages of the Alaska pipeline punctuation were marked by the development and enactment of a federal law that simultaneously settled the Alaska Native land claims and prevented those land claims from obstructing the proposed route of the Alaska pipeline. The Alaska Natives were indigenous peoples who had inhabited Alaska for thousands of years (Fagan 2004). Despite this long history of inhabitation, the question of what lands in Alaska belonged to the Alaska Natives would remain largely unresolved until the Alaska pipeline punctuation. In the eighteenth century, Russia had established a colony in Alaska with an economy driven by an international trade in furs harvested from marine mammals in coastal Alaska (Banner 2007; Black 2004; Bockstoce 2009; Dolin 2010; McBeath and Morehouse 1994). The United States subsequently purchased Alaska from Russia in 1867, but the question of Alaska Native land claims was not resolved either by the Russian colonization of Alaska or by the American purchase of Alaska (Banner 2007; Berry 1975; Boyce and Nilsson 1999; Haycox 2006; McBeath and Morehouse 1994).

An opportunity for Alaska Native groups to seek a federal settlement of their land claims emerged when Alaska became a state in 1959 under the Alaska Statehood Act (Pub. L. 85-508) enacted in the previous year (McBeath and Morehouse 1994; Mitchell 2003; Naske and Slotnick 2011; Whitehead 2004). The Alaska Statehood Act effectively reserved for Congress the authority to settle the Alaska Native land claims (Berry 1975; Boyce and Nilsson 1999; Mitchell 2001). At the time of Alaska statehood, the vast majority of Alaska was owned by the federal government, with private lands in Alaska covering approximately 600,000 acres (H.R. Rep. No. 95-1045 Part I 1978, 65–66, 192; Naske and Slotnick 2011). The Alaska Statehood Act created a process by which the newly established Alaska state government could select, over a twenty-five-year period beginning in 1959, approximately 103 million acres of land from the vast federal domain in Alaska to be transferred into state ownership (Berry 1975, 28; McBeath and Morehouse 1994; Mitchell 2001; Nelson 2004, 36; Ross 2000; S. Rep. No. 95-1300 1978). Additional lands in Alaska were granted to the state to support university and mental health endeavors, bringing the total land entitlement of the state of Alaska to approximately 104 million acres (H.R. Rep. No. 96–97 Part I 1979, 560). The new state of Alaska was thereby entitled to an unprecedented statehood land grant constituting an area larger than the state of California (H.R. Rep. No. 95-1045 Part I 1978, 65–66, 70). The new state of Alaska also received title to 44 million acres of submerged and tidal lands along the Alaska coastline (S. Rep. No. 96-413 1979, 371).

Initial Alaska state land selections focused on lands useful for settlement, highways, and natural resource development. A state land selection that would prove particularly important was in the Prudhoe Bay area on the North Slope of Alaska, where the great Prudhoe Bay oil field would later be found (Coen 2012, 110; Federal Field Committee for Development Planning in Alaska 1968; H.R. Rep. No. 95-1045 Part I 1978, 192–93, 346). However, the state of Alaska did not move quickly to select all of its land entitlement. At the time of Alaska statehood, Alaska lands had not been fully studied, and the new state of Alaska did not yet have the funding needed to adequately manage all the lands it was entitled to select. Furthermore, the twenty-five-year period of land selection appeared at first to give the state of Alaska ample time to study and select available lands for their economic potential and other values. Therefore, the state of Alaska initially pursued a slow process of studying and selecting lands (S. Rep. No. 95-1300 1978, 389; Williss 2005, 33). This gradual approach to state land selection backfired on the state of Alaska when Alaska Native groups and their supporters appealed to the U.S. Department of the Interior to block state land selections until the Alaska Native land claims could be resolved (Boyce and Nilsson 1999; McBeath and Morehouse 1994; Mitchell 2001). The Alaska Native groups organized their land claim efforts by establishing the Alaska Federation of Natives, an organization that would play a prominent role in promoting a land settlement for the Alaska Natives (Berry 1975; Mitchell 2001).

The idea of a land claims settlement for the Alaska Natives found support within the U.S. Department of the Interior. Beginning in 1966, U.S. Secretary of the Interior Stewart Udall pursued a land freeze policy that prevented the leasing and transfer of ownership of federal lands in Alaska. The intent of the land freeze policy was to encourage congressional action to settle the Alaska Native land claims (Allin 2008; Berry 1975; Boyce and Nilsson 1999; Mitchell 2001). This land freeze policy effectively halted state land selections, thereby posing a serious threat to the development aspirations of the state of

Alaska (Boyce and Nilsson 1999; Mitchell 2001). Due to the slow land selection process pursued by the state of Alaska, the state had not yet selected most of the land entitlement available to it under the Alaska Statehood Act at the time of the land freeze (S. Rep. No. 95-1300 1978, 389). A lawsuit against the Alaska land freeze policy was filed in 1967 on the orders of Alaska governor Hickel (Mitchell 2001). A 1968 federal court decision finding against the land freeze was appealed by the Interior Department, ensuring further delays in the courts (Boyce and Nilsson 1999; Mitchell 2001). Governor Hickel then launched a sudden initiative by which the state of Alaska sought to select all of the remaining land area available to the state under the terms of the Alaska Statehood Act (Mitchell 2001). This apparent attempt to circumvent the land freeze was met with a formal but temporary perpetuation of the land freeze by Interior Secretary Udall during his final days in office; in 1969 Secretary Udall signed Public Land Order 4582 to maintain the Alaska land freeze until the end of 1970 (Boyce and Nilsson 1999; Mitchell 2001).

In 1969, Walter Hickel replaced Udall as Interior secretary. Under pressure from Congress, Interior Secretary Hickel proceeded to maintain the land freeze with modifications approved by the House and Senate Committees on Interior and Insular Affairs (Flippen 2000; Mitchell 2001; Nelson 2004). The 1968 federal court finding against the Alaska land freeze was reversed and remanded on appeal in 1969, further delaying the legal challenge to the land freeze (Boyce and Nilsson 1999; Mitchell 2001). While Interior Secretary Hickel would succeed in modifying the land freeze to create a route for the Alaska pipeline project in 1970, subsequent legal actions initiated by five Alaska Native villages nevertheless interrupted the pipeline project (Flippen 2000; Mitchell 2001).

While the Alaska Statehood Act had reserved for the federal government the authority to settle the Alaska Native land claims, there was little progress on a comprehensive Alaska Native land settlement until the Trans-Alaska pipeline proposal in 1969. The Alaska pipeline project therefore encountered the issue of unresolved Alaska Native land claims along the proposed pipeline route. These unresolved land claims created an opportunity for attorneys representing five Alaska Native villages to file a lawsuit in federal court in 1970, asking the court to block pipeline construction across lands claimed by those villages. This legal action represented the response of the five Native villages to the breaking of a promise by an oil industry representative that members of the villages would be given pipeline-related work (Berry 1975; Mitchell 2001). In 1970, a federal judge issued an injunction that blocked the Interior Department from issuing a pipeline construction permit for a section of the Alaska pipeline route crossing lands claimed by one Native village (Berry 1975; Mitchell 2001). This injunction had policy implications far beyond the employment concerns of the five Native villages involved in the lawsuit, as the injunction signaled that Alaska Native land claims held the potential to delay the pipeline project until those land claims were resolved at any place where those land claims intersected the pipeline route (Mitchell 2001; Naske and Slotnick 2011). The injunction therefore provided considerable political leverage for the supporters of a land claims settlement for Alaska Natives.

The oil industry responded to this first injunction against the Trans-Alaska pipeline project by supporting a settlement of the Alaska Native land claims. In 1970, the TAPS oil industry consortium was replaced by the Alyeska Pipeline Service Company (Alyeska), which subsequently endured as the consortium representing the oil corporations active in the Trans-Alaska pipeline project (Alyeska 2009; Coates 1993). The chief executive officer of Alyeska stated in 1970 that a land claims settlement for the Alaska Natives was a precondition for the construction of the Alaska pipeline, and the oil industry began to urge Congress to enact such a settlement (Berry 1975; Mitchell 2001; Willis 2010).

The scale of the Alaska Native land claims settlement became a subject of debate. In 1967, Alaska governor Hickel had appointed a Land Claims Task Force to consider the issue of the Alaska Native land claims, and in 1968, this task force had recommended a land settlement of 40 million acres for the Alaska Natives (Berry 1975, 51, 58; Mitchell 2001, 150–54; Norris 2002, 46). The debate over the Alaska Native lands issue was also informed by a major government study by the Federal Field Committee for Development Planning in Alaska (referred to here as the Federal Field Committee), which was a federally funded planning organization established to provide advice on development in Alaska following a destructive 1964 earthquake in the state. In 1968, the Federal Field Committee issued a report recommending a relatively small land settlement for the Alaska Natives (Federal Field Committee for Development Planning in Alaska 1968; Mitchell 2001, 204–5).

The Alaska Federation of Natives supported a large land settlement proposal in the range of 40 million to 60 million acres (Mitchell 2001; Naske and Slotnick 2011). The idea of such a large-scale land settlement encountered significant opposition in Congress, but nevertheless gained support within the administration of President Richard Nixon; this support reflected the dual goals of President Nixon and members of his administration to both advance the Trans-Alaska pipeline project and advance the cause of Native American rights (Mitchell 2001; Nelson 2004; Nixon 1970). The Nixon administration proposed an Alaska Native land claims settlement of 40 million acres, creating significant political momentum in Congress for this large-scale land settlement (Mitchell 2001; Nixon 1971b).

The debate over an Alaska Native land claims settlement expanded over time into the larger issue of conservation planning in Alaska. The Interior Department had previously drawn up proposals to create large new conservation areas in Alaska, and environmental groups had also previously pursued land conservation efforts in Alaska (Cahn 1982, 9). The Federal Field Committee had also developed an interest in federal conservation planning in Alaska. The 1968 Federal Field Committee report on Alaska Native land claims emphasized the importance of conservation planning in Alaska, stating that "as the last and most extensive wilderness area in the United States and an area of unparalleled grandeur, the development of National Wildlife Ranges and Refuges and National Parks in Alaska is also of high priority" (Federal Field Committee for Development Planning in Alaska 1968, 537). A land conservation planning section was added to a 1970 version of the Alaska Native land claims bill at the suggestion of staff from the Federal Field Committee (Cahn 1982, 11; Mitchell 2001, 441–42; Williss 2005, 35). Although that particular bill was not enacted, environmentalists would subsequently work to link the Alaska Native land claims settlement with conservation planning in Alaska.

In 1971, a number of environmental organizations (including The Wilderness Society, Sierra Club, Friends of the Earth, and others) formed an umbrella organization named the Alaska Coalition to lobby for large-scale land conservation in Alaska (Mitchell 2001, 444; Nelson 2004; Williss 2005, 36). Environmental interests pursuing Alaska lands conservation found allies in Congress. Representatives Morris Udall of Arizona and John Saylor of Pennsylvania introduced an amendment to the Alaska Native land claims bill designed to establish a systematic process of land conservation planning in Alaska (Smith 2006). However, the proposal to combine land conservation efforts into the Alaska Native land claims settlement bill encountered opposition because it held the potential to impede some types of economic development in Alaska. The Nixon administration, state of Alaska, Alaska Federation of Natives, mining interests, logging interests, and oil interests all opposed the inclusion of land conservation efforts into the Alaska Native land claims settlement (Cahn 1982, 12). But these arguments against land conservation in Alaska were undermined by the massive economic potential of the Trans-Alaska pipeline project that the land claims settlement bill was intended to enable, as the oil production made possible by the pipeline would give the economy of Alaska excellent prospects even if large areas of Alaska were conserved (Mitchell 2001).

The Udall-Saylor amendment for Alaska lands conservation planning failed a vote in the House of Representatives, but environmental interests and their allies continued to pursue a conservation planning amendment to the Alaska Native land claims bill in the Senate (Williss 2005, 36–40). Leaders from the environmental movement and the National Park Service lobbied key senators on this issue, focusing particularly on Senator Alan Bible of Nevada, who was chairman of the Senate Subcommittee on Parks and Recreation. National Park Service director George Hartzog invited Senator Bible on an extensive tour of the areas proposed for protection in Alaska. Hartzog also invited Sierra Club leaders Edgar and Peggy Wayburn to join this trip, thereby giving leaders from the National Park Service and the environmental movement a significant opportunity to jointly lobby Senator Bible for conservation in Alaska (Hartzog 1988; Wayburn and Alsup 2004; Williss 2005). Senator Bible subsequently referred to this trip in the Senate as he proposed an amendment to the Alaska Native land claims bill that would create a program designed to temporarily reserve large areas of Alaska for conservation planning purposes (Elliott 1994; *U.S. Congressional Record* 1971, 38451-53). The Alaska conservation planning amendment introduced by Senator Bible passed a Senate vote and was further modified in the subsequent conference committee (Cahn 1982, 12). The final Alaska Native land claims bill contained sections that authorized a systematic process for the Interior secretary to set aside vast areas of federal lands in Alaska for consideration as conservation areas (Allin 2008; Nelson 2004). The final bill also set up a land selection formula that would distribute land and funding not only among Alaska Native individuals but also among Native corporations to be established at both the village and regional levels in Alaska. The land selection formula in the final bill resulted in a land settlement of 44 million acres for Alaska Natives (Mitchell 2001).

The Alaska Native Claims Settlement Act of 1971 (Pub. L. 92-203) settled the Alaska Native land claims by allowing Alaska Natives to select 44 million acres of federal lands in Alaska for transfer into the ownership of Alaska Native village corporations, Alaska Native regional corporations, and individual Alaska Natives (Mitchell 2001). The Alaska Native Claims Settlement Act also distributed $962.5 million in funding to the Alaska Natives (Berry 1975; Ross 2000). The Alaska Native Claims Settlement Act prevented both the Alaska Natives and the state of Alaska from selecting lands within a transportation and utility corridor—the type of corridor needed for the Alaska pipeline route—if such a corridor was established by the Interior secretary (85

Stat. 708). Therefore, the Alaska Native Claims Settlement Act simultane-ously settled the Alaska Native land claims and created a method to clear the route for the Alaska pipeline (Berry 1975; Coates 1993).

This element of the Alaska pipeline punctuation was critical because it both resolved the issue of indigenous land claims in Alaska and resolved the implications of those claims for the Alaska pipeline project. But the Alaska Native land claims settlement did not resolve the debate over the environ-mental impacts of the proposed pipeline project, a debate that would culmi-nate in the approval of another federal law that marked the end of the Alaska pipeline punctuation. Therefore, the establishment of the Alaska Native Claims Settlement Act as an institutional arrangement to clear the pipeline route marked the early stages—but not the conclusion—of the critical period that allowed the building of the Alaska pipeline.

National Environmental Policy and the Trans-Alaska Pipeline

Legal actions interfering with the Alaska pipeline project were pursued not only by Alaska Native villages but also by national environmental groups that hoped to stop the pipeline project or at least mitigate the impacts of the pipeline project on the environment (Mitchell 2001; Nelson 2004). A key foundation of the legal challenges to the Alaska pipeline proposal by environ-mental groups was the National Environmental Policy Act of 1969 (Pub. L. 91-190), a law that had been enacted in the same year that the Alaska pipeline project was proposed. The National Environmental Policy Act required the completion of an environmental impact statement to consider (1) the potential environmental impacts of proposed major federal actions significantly affecting the environment and (2) alternatives to those proposed actions (Ashenmiller 2006; Cicchetti 1972; Coates 1993). The Interior Department took the lead in drafting the environmental impact statements related to the Alaska pipeline proposal, since most of the pipeline route was on federal lands. In 1970, three environmental groups filed a legal action challenging the Interior Department's permit for a road meant to support the Alaska pipeline project; the Interior Department had produced an eight-page environmental impact statement for this road, but no such statement for the proposed Alaska pipeline. The environmental groups The Wilderness Society, Friends of the Earth, and Environmental Defense Fund claimed in court that this brief environmental impact statement for the road failed to satisfy the requirements of the National Environmental Policy Act for the Alaska pipeline project. The environmental groups also claimed in court that the Alaska pipeline project would exceed the federal limits on pipeline right-of-way widths set by the Mineral Leasing Act of 1920 (Pub. L. 66-146), a law that had established wide-reaching federal regulations over oil development on federal lands (Ashenmiller 2006; Berry 1975, 118; Coates 1993, 189–90; Mitchell 2001; Ross 2000; Worster 1994). In 1970, a federal judge found that an environmental impact statement for the entire Alaska pipeline project (not just the supporting road) was required by the National Environmental Policy Act and also found that the pipeline project violated the right-of-way width requirements of the Mineral Leasing Act. The judge issued a preliminary injunction blocking construction of the Alaska pipeline (Berry 1975; Coates 1993). Due to this injunction, the conflicts over both the environmental impact statement and the right-of-way width for the Alaska pipeline project became significant obstacles to that project.

The environmental debate over the Alaska pipeline proposal continued well into 1973. The development of an environmental impact statement for the Alaska pipeline project proved controversial, and both the draft and final versions of that statement were met with substantial criticism (Ashenmiller 2006; Coates 1993; Nelson 2004; The Wilderness Society, Environmental Defense Fund, and Friends of the Earth 1972; U.S. Department of the Interior 1971, 1972). The Interior Department held public hearings on the draft environmental impact statement and solicited comments on the draft statement from other federal agencies and the Canadian government (Canada's interest in the draft statement was not only in the possibility of a Trans-Canada pipeline but also in the risk of oil spills from tankers that would transport North Slope oil along the west coast of Canada if the Trans-Alaska pipeline was built). The Interior Department received a large number of written and oral comments on the draft environmental impact statement; most of the recurring comments were critical of that statement. In addition, the government of Canada and the newly founded U.S. Environmental Protection Agency separately expressed significant concerns over the potential environmental impacts of the proposed Trans-Alaska Pipeline System (U.S. Department of the Interior 1972).

In 1972, the Interior Department released the six-volume final environmental impact statement for the Alaska pipeline proposal (U.S. Department of the Interior 1972). Interior Secretary Rogers Morton (who had replaced Hickel in that position) announced a forty-five-day period for comments on the final environmental impact statement, but no public hearings on that statement were held. The final environmental impact statement indicated that the Alaska pipeline project posed significant hazards to the environment, including the possibility of marine oil spills (U.S. Department of the Interior 1972). A group of environmental organizations responded to the final environmental impact statement for the Alaska pipeline proposal by compiling a four-volume collection of negative comments on the statement written by a number of different individuals and organizations (The Wilderness Society, Environmental Defense Fund, and Friends of the Earth 1972). In this counterstatement, the environmental organizations attempted to counteract the considerable governmental expertise applied in developing the environmental impact statement by drawing on independent experts for comments on that statement (including a number of researchers working at universities around the nation). Interior Secretary Morton nevertheless decided to issue the right-of-way permit for the Alaska pipeline in 1972 (Coates 1993; Siehl and Want 1972; The Wilderness Society, Environmental Defense Fund, and Friends of the Earth 1972).

The completion of the final environmental impact statement for the proposed Alaska pipeline led a federal judge to dissolve the preliminary injunction against the Alaska pipeline project. A group of environmental organizations continued to pursue further legal action against the Alaska pipeline project, and in 1973 the pipeline project was blocked again when a U.S. District Court found that the pipeline proposal violated the right-of-way width requirements of the Mineral Leasing Act (H.R. Rep. No. 93-414 1973, 9). Therefore, the final environmental impact statement for the proposed Alaska pipeline did not resolve the conflict over the pipeline project (Worster 1994).

The Alaska pipeline debate was finally resolved in Congress. At issue were changes to the Mineral Leasing Act, which contained a pipeline right-of-way width limit that allowed insufficient room for modern construction machinery to maneuver (Coates 1993; Worster 1994). A revision of this archaic right-of-way width limit was needed to allow the use of modern construction machinery in the building of the Alaska pipeline. A 1973 House of Representatives report noted that the strict court interpretation of the right-of-way width limits for pipelines under the Mineral Leasing Act "effectively precludes construction of any major pipeline on public lands" (H.R. Rep. No. 93-414 1973, 9). Also at issue was the continuing debate over the adequacy of the final environmental impact statement for the Alaska pipeline, which constituted a controversial element of the congressional debate over a bill designed to authorize the Alaska pipeline. Senator Mike Gravel of Alaska proposed an amendment to the pipeline bill that would exempt the Alaska pipeline from further action under the National Environmental Policy Act (Ross 2000).

This proposed Gravel amendment proved controversial in the Senate, with Senator Henry Jackson of Washington protesting that the Gravel amendment "sets a dangerous precedent by attempting to exempt a project which all parties admit represents a potential threat to the environment from the requirements of environmental law" (*U.S. Congressional Record* 1973, 24301). The debate over the Gravel amendment produced a tie vote in the Senate; a tie-breaking vote cast by Vice President Spiro Agnew allowed passage of the Gravel amendment in the Senate (Coates 1993; *U.S. Congressional Record* 1973, 24316). In the end, Congress removed the two key obstacles to the Alaska pipeline by passing a bill that (1) disallowed further action under the National Environmental Policy Act concerning the Alaska pipeline proposal and (2) allowed exceptions to the right-of-way width limit specified in the Mineral Leasing Act (Ross 2000; Worster 1994).

An important factor driving this critical period of reform was the rapid rise in oil prices that occurred during this period. These rising oil prices dramatically increased the potential economic value of the Alaska pipeline project. Oil prices more than doubled between December 1970 and September 1973, and in October 1973, some members of the Organization of Petroleum Exporting Countries (OPEC) decided to place an embargo on oil exports to the United States and unilaterally increase oil prices further (Maugeri 2006). The rising oil prices in this period strengthened the business case for the Alaska pipeline on the part of the oil industry, while the combination of escalating oil prices and the OPEC embargo strengthened the political case for approval of the Alaska pipeline on the part of the federal government (Flippen 2000). The impact of the energy crisis in strengthening the political case for the Alaska pipeline is evident both in the congressional debate over the Alaska pipeline and in statements from President Nixon (Nixon 1973a, 1973b, 1973c; *U.S. Congressional Record* 1973). Senator Ted Stevens of Alaska stated that the availability of North Slope oil "at the earliest possible time is of critical national importance in view of the increasing crisis of energy supply facing the Nation over the next decade" (*U.S. Congressional Record* 1973, 24305). As the energy crisis mounted, Nixon focused attention on the Alaska pipeline proposal as a means to increase secure domestic oil production and reduce American dependence on insecure foreign sources of oil. In 1973, Nixon issued a series of statements urging congressional action to approve the Alaska pipeline (Nixon 1973a, 1973b, 1973c). In a message to Congress on November 8, 1973, Nixon stated that "legislation authorizing construction of the Alaskan pipeline must be the first order of business as we tackle our long-range energy problems" (Nixon 1973c, 925). The Alaska pipeline bill passed by overwhelming majorities in Congress and was signed into law by

Nixon on November 16, 1973 (Ashenmiller 2006; Coates 1993). In a statement issued on the day that he signed the Alaska pipeline bill, Nixon stated that the bill was "the first piece of major legislation by this Congress dealing with our energy requirements" and also stated that "we must take every available step to utilize our vast, but untapped, domestic energy resources in order to avoid dependence on foreign sources of supply in the long run" (Nixon 1973d, 945).

The Alaska pipeline bill was introduced as the Federal Lands Right-of-Way Act but would become more generally known as the 1973 Trans-Alaska Pipeline Authorization Act (Pub. L. 93-153). Title I of the Trans-Alaska Pipeline Authorization Act authorized the Interior secretary and the heads of other federal agencies to widen the right-of-way for a pipeline beyond the limits set by the Mineral Leasing Act of 1920. This reform allowed the right-of-way for pipelines on federal lands to be increased to any extent necessary for pipeline operations and also allowed the granting of temporary permits for the use of additional federal lands for pipeline construction. This reform thereby removed the basis for further legal challenges to the Alaska pipeline project under the Mineral Leasing Act. Title II of the Trans-Alaska Pipeline Authorization Act indicated the intent of Congress to expedite the construction of the pipeline, stating that "the early development and delivery of oil and gas from Alaska's North Slope to domestic markets is in the national interest because of growing domestic shortages and increasing dependence upon insecure foreign sources" (87 Stat. 584). Title II further stated that "the purpose of this title is to insure that, because of the extensive governmental studies already made of this project and the national interest in early delivery of North Slope oil to domestic markets, the trans-Alaska oil pipeline be constructed promptly without further administrative or judicial delay or impediment" (87 Stat. 584). The Trans-Alaska Pipeline Authorization Act declared that no further action would be taken under the National Environmental Policy Act concerning the construction and operation of the Trans-Alaska pipeline, thereby circumventing the debate over the environmental impact statement for the pipeline project (87 Stat. 585). The approval of the Trans-Alaska Pipeline Authorization Act therefore marked the end of the Alaska pipeline punctuation, as this law was the final institutional arrangement necessary for the building of the Alaska pipeline. The injunction blocking the Alaska pipeline project was subsequently dissolved in a federal court in 1974 (Coates 1993).

In a 1971 message to Congress, Nixon had proposed the precursor to the policy image that would prevail in the Alaska pipeline punctuation. In this message Nixon proposed a comprehensive approach to federal land use planning in Alaska that "should take account of the needs and aspirations of the native peoples, the importance of balanced economic development, and the special need for maintaining and protecting the unique natural heritage of Alaska. This can be accomplished through a system of parks, wilderness, recreation, and wildlife areas" (Nixon 1971a, 135). Nixon's central effort for economic development in Alaska was the Trans-Alaska pipeline project (Nixon 1973a, 1973b, 1973c, 1973d). In the end, the Alaska pipeline punctuation established a multifaceted policy image containing the three major elements outlined in Nixon's statements. The first element was stated in the Alaska Native Claims Settlement Act as the "immediate need for a fair and just settlement of all claims by Natives and Native groups of Alaska, based on aboriginal land claims" (85 Stat. 688) with additional provisions that prevented those claims from delaying the Alaska pipeline (85 Stat. 708). The second element was stated in the Trans-Alaska Pipeline Authorization Act as supporting the "earliest possible construction of a trans-Alaska oil pipeline" (87 Stat. 584). The third element was stated by Senator Alan Bible as he proposed a conservation amendment to the Alaska Native Claims Settlement Act "to insure that areas suitable for inclusion in the national park or wildlife refuge systems are protected for a reasonable period of time so that Congress may consider legislation on this subject" (Elliott 1994; *U.S. Congressional Record* 1971, 38452). The critical period examined in this study thereby established a multifaceted policy image that supported (1) a large land and financial settlement for the Alaska Natives whose land claims might otherwise have interfered with the pipeline route, (2) the building of the Alaska pipeline to allow large-scale oil development in northern Alaska, and (3) a systematic program for the temporary protection of vast areas of Alaska to allow consideration of federal conservation options in those areas (McBeath and Morehouse 1994). This multifaceted policy image provided the supporting ideas for the Alaska Native Claims Settlement Act and the Trans-Alaska Pipeline Authorization Act, which together constituted the two defining institutional arrangements established during the Alaska pipeline punctuation.

The authorization of the Alaska pipeline reflected a pattern of predominantly supportive political and media attention to oil development in Alaska during the critical period. These patterns of attention are examined further in the following section.

Congressional and Media Attention

The punctuated equilibrium theory identifies political and media attention as key independent variables driving the dependent variable of policy reform in punctuations (Baumgartner and Jones 2009). A rise in political and media attention in favor of a proposed policy reform (measured through the coding of legislative hearings and media reports) can drive punctuations (measured by the enactment of new institutional arrangements and supporting policy images). This study finds evidence of such a rise in political and media attention in favor of oil development in Alaska during the period of the Alaska pipeline debate. Summaries of all congressional hearings and *New York Times* articles on the topic of oil in Alaska were collected and coded for this study, covering the period 1959–80 (beginning with Alaska statehood and encompassing the period during which the Alaska pipeline was debated and eventually built). To assess patterns of support for oil development in Alaska over time, the topics in these congressional hearings and news articles were coded positive or negative in tone toward oil development in Alaska. Topics coded positive were supportive of oil development in Alaska, while topics coded negative were not supportive of oil development in Alaska. These data sources and coding procedures are typical of empirical research on the punctuated equilibrium theory (Baumgartner and Jones 2009).

The Alaska pipeline punctuation was accompanied by a sharp increase in congressional and media attention in support of oil development in Alaska. Figure 2.1 shows the annual number of congressional hearing topics coded positive or negative in tone toward oil development in Alaska in the period 1959–80. Because some hearings included multiple topics, the number of topics coded is larger than the number of hearings.

The early years of this record include an extended period (1960–68) with no attention to oil in Alaska in congressional hearings. The Alaska pipeline punctuation was marked by a sharp rise in positive tone topics supportive of Alaska oil development discussed in congressional hearings both in 1969 (the year of the Alaska pipeline proposal) and in 1973 (the year that the Alaska pipeline was authorized by Congress). Overall, the number of positive tone topics outweighed the number of negative tone topics in the 1969 and 1973 congressional hearings on oil in Alaska. Yet this wave of predominantly positive attention was short-lived. In several years following the end of the Alaska pipeline punctuation in 1973, the number of negative tone topics outweighed the number of positive tone topics in congressional hearings on oil in Alaska. This rise in negative attention reflected substantial congressional concerns over the environmental safety of the pipeline project (H.R. Rep. No. 95-5 1977). The Alaska pipeline punctuation was therefore accompanied by a rise in positive political attention to oil development in Alaska in the years 1969–73. The institutional arrangements established in this punctuation subsequently endured and allowed the construction of the Alaska pipeline despite often negative congressional attention to oil development in Alaska in the years 1974–80.

FIGURE 2.1. Topics in U.S. congressional hearings on Alaska and oil (1959–80) coded positive or negative in tone toward oil development in Alaska. Topics coded positive were supportive of oil development in Alaska; topics coded negative were not supportive of oil development in Alaska.

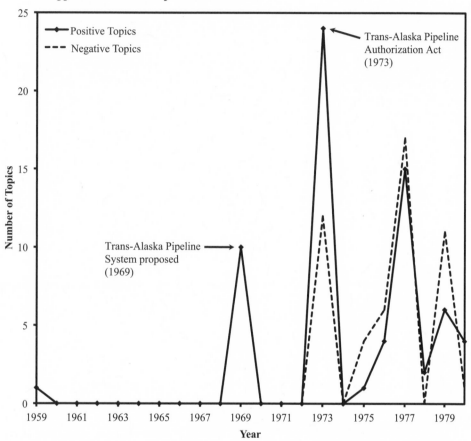

Figure from Busenberg (2011). © The Policy Studies Organization. Reprinted by permission.

Positive attention to oil development in Alaska during the Alaska pipeline punctuation is evident not only in data from congressional hearings but also in data from newspaper articles in the *New York Times*. Figure 2.2 shows the annual number of *New York Times* article topics coded positive or negative in tone toward oil development in Alaska in the period 1959–80. There is a notable difference between the congressional hearings data and the *New York Times* data concerning oil in Alaska; while the congressional hearings often

FIGURE 2.2. Topics in *New York Times* articles on Alaska and oil (1959–80) coded positive or negative in tone toward oil development in Alaska. Article topics coded positive were supportive of oil development in Alaska; article topics coded negative were not supportive of oil development in Alaska.

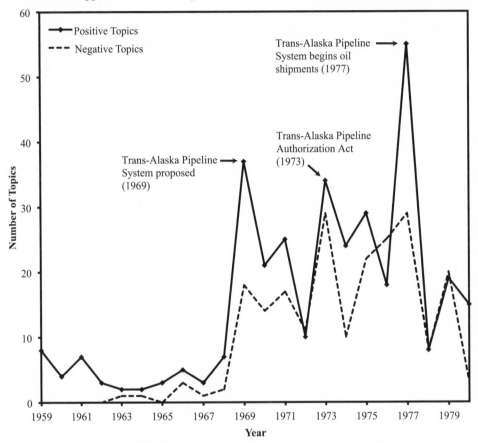

Figure from Busenberg (2011). © The Policy Studies Organization. Reprinted by permission.

addressed multiple major topics per hearing, the *New York Times* articles generally addressed one major topic per article. Therefore, one major topic concerning oil in Alaska was coded for each *New York Times* article. This approach to coding topics in congressional hearings data and *New York Times* data is used throughout this book.

The first years of this media record (1959–68) marked a period of limited attention to oil in Alaska in the *New York Times*. The Alaska pipeline punctuation was characterized by a sharp rise in positive tone article topics supporting oil development in Alaska, with peaks of positive attention in 1969 and 1973. Overall, the number of positive tone article topics outweighed the number of negative tone article topics during the critical period. In addition, a prevailing positive tone was evident in *New York Times* article topics on oil in Alaska for most of the years in the data set. Positive attention to oil development in Alaska in *New York Times* article topics peaked in 1977, the year in which the Trans-Alaska Pipeline System began oil shipments.

Overall, the record of congressional hearings and *New York Times* articles indicates a sharp rise in political and media attention that was predominantly supportive of oil development in Alaska during the Alaska pipeline punctuation. These findings support a central concept in the punctuated equilibrium theory in the context of this case; punctuations are driven by a rise in political and media attention in support of reform in a policy domain (Baumgartner and Jones 2009).

Critical periods of policy reform are particularly consequential because they can create enduring institutional arrangements with enduring policy consequences. The enduring policy consequences of the critical period that authorized the Alaska pipeline project are considered in the following sections, beginning with an examination of the decision made during this critical period to reject a Trans-Canada pipeline as an alternative to the Trans-Alaska pipeline.

The Trans-Canada Pipeline Alternative

The enactment of the Trans-Alaska Pipeline Authorization Act in 1973 marked a major defeat for the national environmental groups that had sought to block or at least reroute the pipeline. The Alaska pipeline punctuation allowed the building of an oil transportation system that posed an enduring threat of marine oil spills. An alternative pipeline proposal that did not entail the risk of marine oil spills was defeated in the course of this critical period; the proposed Trans-Canada pipeline would have transported North Slope oil across both Alaska and Canada to the American market by pipeline alone, thereby eliminating the need for the hazardous marine oil transportation system that was an integral part of the Trans-Alaska Pipeline System. The Trans-Canada pipeline proposal received some support from both Congress and Canada (Berry 1975; Kirkey 1997). Canada stood to benefit economically from the construction and operation of a Trans-Canada pipeline (as did some U.S. states that would receive oil from such a pipeline). Furthermore, the

Canadian government was concerned that the marine oil shipments associated with the Trans-Alaska route could lead to oil spills affecting the west coast of Canada (a problem that the Trans-Canada route would avoid).

A comparison between the two routes revealed both environmental benefits and drawbacks to the Trans-Canada route. Because the Trans-Canada route was considerably longer than the Trans-Alaska route, a Trans-Canada pipeline would have caused more extensive environmental damage during construction (and exposed a larger area of land to the risk of oil spills in operation) when compared to a Trans-Alaska pipeline. However, the Trans-Canada pipeline proposal also offered two major environmental benefits when compared to the Trans-Alaska pipeline proposal. First, oil pipelines had a better environmental safety record than oil tankers (and the Trans-Canada pipeline would not make use of tankers). Second, the Trans-Canada pipeline route would avoid some of the earthquake hazards found along the Trans-Alaska pipeline route. On balance, the final environmental impact statement for the Trans-Alaska pipeline proposal gave substantial support to the argument that the Trans-Canada pipeline would cause less overall environmental damage than the Trans-Alaska pipeline (Brew 1974; Cicchetti 1972; U.S. Department of the Interior 1972). By 1973, the coalition of environmental organizations that objected to the Alaska pipeline proposal had concluded that a system to transport oil from Alaska's North Slope would be approved. In 1973, the members of this environmentalist coalition announced their collective support for the Trans-Canada pipeline as an alternative posing lesser risks to the environment than the Trans-Alaska pipeline (Berry 1975, 236; Coates 1993, 241; S. Rep. No. 93-207 1973, 19; The Wilderness Society, Environmental Defense Fund, and Friends of the Earth 1972, 5).

The Trans-Canada pipeline proposal was complicated by the likelihood of delays, additional regulation, and additional taxation when compared to the Trans-Alaska pipeline proposal. The viability of the Trans-Canada route depended on the approval of the Canadian government, and the prolonged environmental impact assessment process for the Trans-Alaska pipeline project suggested that the approval process for a Trans-Canada pipeline project might also involve substantial delays. Furthermore, the Trans-Canada route would make the pipeline project the subject of regulation and taxation in Canada as well as in the United States, raising the possibility of greater taxation and regulatory complexity than the Trans-Alaska route (Berry 1975; Cicchetti 1972; Cicchetti and Freeman 1973; Kirkey 1997).

The oil industry and the Nixon administration criticized the Trans-Canada route as involving a greater likelihood of delays and uncertainties when compared to the Trans-Alaska route (Alyeska n.d.; Morton 1972; Nixon 1973a).

In a statement announcing his decision to issue a right-of-way permit for the Trans-Alaska pipeline, Interior Secretary Morton noted that the Trans-Alaska and Trans-Canada routes were the two alternatives to which he had given the most serious consideration (Morton 1972). In this statement Morton argued that a bilateral arrangement for a Trans-Canada pipeline was impractical due in part to the "delay of project pending the completion of environmental, engineering, and construction studies for a Canadian route" and also argued that "it is our best national interest to avoid all further delays and uncertainties in planning the development of Alaska North Slope oil reserves by having a secure pipeline located under the total jurisdiction and for the exclusive use of the United States" (Morton 1972, 4). The energy crisis had created political pressure for a Trans-Alaska pipeline that could begin transporting oil sooner than the Trans-Canada pipeline alternative. In a 1973 message to Congress concerning the energy challenges facing America, Nixon stated his view on the two alternatives for North Slope oil transportation: "I oppose any further delay in order to restudy the advisability of building the pipeline through Canada. Our interest in rapidly increasing our supply of oil is best served by an Alaskan pipeline. It could be completed much more quickly than a Canadian pipeline" (Nixon 1973a, 307).

Concerns over the potential delays of a Trans-Canada route were also evident in congressional debates on this issue, with Senator Henry Jackson stating his concern over "the inevitable delay in designing, organizing, and getting authorization for a new, international pipeline at a time when a wholly domestic project is ready to go" (*U.S. Congressional Record* 1973, 24314). A 1973 Senate report noted the extensive preparations already made for the Alaska pipeline project (referred to in the report as the Alyeska project): "The necessary business organization, financial arrangements, engineering design and logistical preparations for the Alyeska project have been completed, so that construction could begin as soon as a right-of-way is granted, while *none* of these necessary preparations has been accomplished for a Trans-Canada route" (S. Rep. No. 93-207 1973, 21). The same Senate report further noted that "a new pipeline route through Canada would, of course, require a new environmental impact statement and public hearings, and involves the possibility of a new round of litigation within the United States" (S. Rep. No. 93-207 1973, 22). The extensive studies and preparations made by both industry and government concerning the Alaska pipeline project were therefore used as an argument for choosing the Trans-Alaska route over the Trans-Canada route. The prospect of a North Slope oil transportation system that could be constructed promptly, make use of proven technologies, and proceed under complete American jurisdiction was politically and commercially attractive to the development and government interests backing the Trans-Alaska route.

The prevailing congressional perspective on the choice between the two pipeline proposals was reflected in the Trans-Alaska Pipeline Authorization Act, which stated that a Trans-Canada pipeline "may be needed later and it should be studied now, but it should not be regarded as an alternative for a trans-Alaska pipeline that does not traverse a foreign country" (87 Stat. 584). The Trans-Alaska Pipeline Authorization Act authorized and directed the Interior secretary to investigate the feasibility of one or more Trans-Canada pipelines for oil or natural gas that could be built in addition to (rather than as a substitute for) the Trans-Alaska pipeline and also authorized and requested the U.S. president to enter into negotiations with Canada to determine the possibilities for oil and natural gas transportation across Canada to America (87 Stat. 588–589). Therefore, Congress chose to authorize the Trans-Alaska pipeline while simultaneously reserving the concept of a Trans-Canada pipeline as a subject for further study and international negotiation. A Trans-Canada oil or gas pipeline from Alaska to America has yet to be built. However, the state of Alaska continues to pursue efforts to build a gas pipeline that would allow the large-scale development of the vast natural gas reserves of the North Slope of Alaska.

The Trans-Alaska Pipeline System and Environmental Protection in Alaska

The choice of the Trans-Alaska pipeline route over the Trans-Canada pipeline route removed the option of transporting North Slope oil by pipeline alone. To reach points outside of Alaska without crossing Canada, the Trans-Alaska pipeline route would connect to a marine oil terminal in Alaska. The port of Valdez in the Prince William Sound region of Alaska was chosen as the site of the marine oil terminal, as this seaport was thought to pose lesser ice hazards to oil shipping than the seas near the North Slope region. The marine oil terminal in Prince William Sound would load oil from the pipeline onto oil tanker vessels for shipment to other regions (Busenberg 2008). Map 2.1 shows the route of the Trans-Alaska pipeline and the location of the Valdez marine oil terminal.

The Trans-Alaska Pipeline System would pose significant hazards to the environment on land and at sea. On land, the oil pipeline would be vulnerable

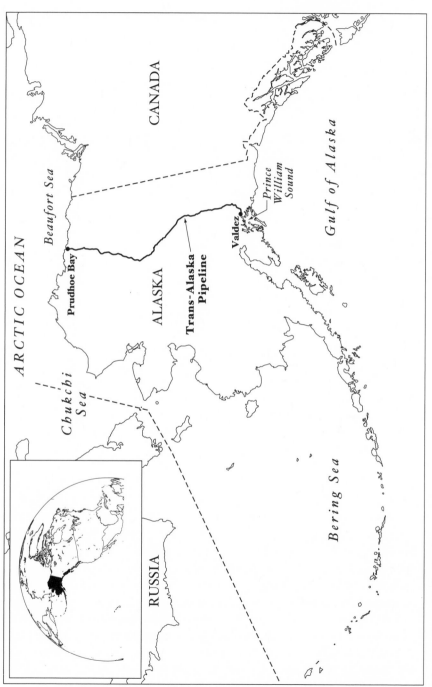

Map 2.1. Trans-Alaska Pipeline.

to earthquakes. The pipeline would also be vulnerable to oil spills as the result of accidents, corrosion, and sabotage. Oil spills from the pipeline could affect land, rivers, and streams along the pipeline route. However, the pipeline project incorporated safeguards against oil pollution. The pipeline was built to resist damage from major earthquakes and was also equipped with leak detection systems and a number of valves that could be used to stop the flow of oil in the pipeline in the event of a pipeline rupture (Alyeska 2009; Coates 1993; McBeath et al. 2008). A road was built next to the pipeline, so both personnel and response equipment could be sent in along that road in the event of a pipeline rupture (Coates 1993). The initial design for a buried Alaska pipeline was revised, as a buried pipeline posed the risk of transferring enough heat from the oil to thaw and destabilize the permafrost soils found along much of the route; this thawing process could have ruptured the pipeline. The pipeline was redesigned to manage the special problems of permafrost. Much of the redesigned pipeline was elevated above the ground on supports designed to transfer heat away from the frozen soils underneath (Coates 1993). The elevated sections of the pipeline were also designed to accommodate thermal expansion and contraction from temperature variations through a zigzag pipeline configuration that allowed the pipeline a range of motion to absorb stresses from thermal effects while maintaining the integrity of the pipeline. This zigzag pattern increased the length of the pipeline from the initial estimates of 789 miles to its actual length of approximately 800 miles. The pipeline was also insulated to help prevent the oil within the pipeline from freezing when oil pumping stopped in the pipeline (Coates 1993, 256–58).

The use of oil tankers in the Trans-Alaska Pipeline System created major threats to the environment beyond those threats posed by the pipeline. Oil tankers in transit could hit coastlines or undersea rocks (groundings) and could also hit other vessels (collisions). Oil tanker collisions and groundings could occur as the result of navigational errors, hazardous weather, loss of vessel propulsion, loss of vessel steering, fires, explosions, or a combination of those factors. Oil tanker accidents could breach the very large cargo tanks of the ships and cause uncontrolled releases of oil into water; the oil would then often disperse into widespread slicks that could be difficult or impossible to contain or recover (Burger 1997). Despite these dangers, the marine oil transportation component of the Trans-Alaska Pipeline System would operate for many years with few safeguards against the inherent hazards of accidental oil spills from tankers (Birkland 1997; Busenberg 2008).

The Trans-Alaska Pipeline System did include a safeguard against the problem of *intentional* oil pollution from tankers. The Trans-Alaska Pipeline System was built at a time when oil tankers around the world intentionally discharged large volumes of oil into the sea as a matter of standard practice. This intentional oil pollution was the result of ballast water exchange, a process in which an oil tanker would pump seawater into and out of its cargo tanks to adjust the weight of the vessel. The weight of oil tankers needed adjusting because unloaded tankers had stability problems in the water. When the cargo tanks of a tanker were loaded full of oil for transportation, the weight of the oil made the vessel ride lower in the water and ensured navigational stability. But when the oil cargo was pumped out of the cargo tanks at the point of delivery, the vessel would rise high enough in the water to cause vessel stability issues. The solution was to fill some of the cargo tanks with seawater; this ballasting operation lowered the tanker in the water, allowing the tanker to maintain stability when not transporting oil cargo. The ballast water would be held in the cargo tanks until the tanker was preparing to accept the next load of oil cargo, at which point the ballast water would be pumped out of the cargo tanks so the oil could be subsequently pumped in. But a significant amount of residual oil remained in the cargo tanks of oil tankers after they were unloaded, and that residual oil would mix with the water pumped into the same cargo tanks for ballast. When the tanker pumped out the ballast water, it would also pump out residual oil mixed in with the ballast water. Oil tankers often pumped out oily ballast water directly into the ocean, causing intentional marine oil pollution (Mitchell 1994).

A long series of attempts to address the problem of intentional oil pollution at sea eventually led to an international agreement that required large new oil tankers to be built with segregated ballast tanks that would never be used to store oil (a readily enforceable design standard that would eliminate the problem of oily ballast water discharge from these ballast tanks). This international agreement for segregated ballast tanks was contained in the 1973 International Convention for the Prevention of Pollution from Ships and a 1978 protocol modifying this convention; together these international agreements are known as MARPOL 73/78 (Mitchell 1994, 70–102). The problem of oily ballast water discharge was nevertheless important in the marine oil trade of Alaska because older oil tankers built without segregated ballast tanks would remain active in that trade for many years (Alaska Oil Spill Commission 1990, 123). The marine oil terminal for the Trans-Alaska Pipeline System was designed to deal with the problem of oil residues in ballast water; the terminal included a Ballast Water Treatment Facility that would remove most of the oil residues from the ballast water of oil tankers and then pump the treated water back into the port of Valdez (Coates 1993, 262; U.S. Department of the Interior 1972, 32).

The location of the marine oil terminal for the Trans-Alaska Pipeline System presented a range of hazards, as the port of Valdez was vulnerable to extremely powerful earthquakes and tsunamis. These hazards were clearly demonstrated by a 1964 earthquake and tsunami event in Prince William Sound. The 1964 earthquake triggered a submarine landslide along the shoreline of the city of Valdez, causing the city's waterfront and docks to slide into the sea. This submarine landslide displaced a mass of seawater and generated a tsunami that caused further damage in Valdez. The disaster set fuel tanks on fire, disabled the city's water system, and caused major soil disturbances. The ground underlying Valdez was subsequently judged by the state of Alaska to be too unstable for the safety of dwellings, so that land was condemned. A new city of Valdez was built on a site four miles to the west, at a higher elevation and on more stable soils than the old city of Valdez. The abandoned buildings of the old city of Valdez were subsequently burned by the state, and the condemned site of old Valdez was used to store pipe sections that would eventually be used to build the Trans-Alaska pipeline (Mason et al. 1997, 56–57, 215–17; Rennick 1993, 53, 61).

The great 1964 earthquake in Alaska was not forgotten as the marine oil terminal in the port of Valdez was designed. The terminal was built on a large area of bedrock that would provide a relatively stable foundation in the event of a major earthquake. A particular concern of the design of the terminal was the presence of large storage tanks at the terminal that would hold the oil from the Trans-Alaska pipeline until that oil could be transferred to the tanker vessels. The oil storage tanks at the terminal were reinforced to resist earthquake damage and were also built on high ground, above the expected zone of tsunami inundation (U.S. Department of the Interior 1972, 31). While the oil terminal at Valdez was located on a site and built in a manner that made the terminal partially resistant to earthquake and tsunami damage, the loading of oil onto tankers was designed to occur near sea level. Therefore, oil tanker operations at the terminal would remain inherently vulnerable to tsunamis. A powerful tsunami could carry large oil tankers docked at the terminal into some of the terminal's infrastructure and the nearby shoreline. A tsunami could therefore wreck both the tankers and part of the terminal's infrastructure. Tsunami-induced destruction, explosions, fires, and catastrophic oil spills therefore remained a risk at the Valdez oil terminal.

The Trans-Alaska pipeline and the associated marine oil transportation system in Prince William Sound became fully operational in 1977 (Coates 1993; McBeath et al. 2008). The building of the Trans-Alaska Pipeline System made possible not only the development of the Prudhoe Bay oil field but also the development of the Kuparuk field and a number of other oil fields scattered across the North Slope of Alaska (McBeath et al. 2008; Worster 1994). All of the oil produced by the various fields on the North Slope of Alaska is transported to the market by the Trans-Alaska pipeline (McBeath et al. 2008). The Alaska pipeline punctuation therefore brought a vast industrial transformation to northern Alaska. Before 1968, the North Slope formed part of a region of northern Alaska that was considered the largest intact area of wild land in the United States (National Research Council 2003). As a result of subsequent oil development, the North Slope of Alaska now contains one of the largest industrial complexes in the world. The North Slope industrial complex of roads, buildings, pipelines, drilling sites, and other infrastructures together extend across an estimated area of one thousand square miles (National Research Council 2003). Taken together, the North Slope industrial complex and the Trans-Alaska Pipeline System constitute a large-scale disturbance of the Alaskan landscape and a continuing hazard to the environment across large areas of Alaska. Throughout more than three decades of operation, the Trans-Alaska Pipeline System and the North Slope oil fields have caused significant environmental damage not only through construction and maintenance activities but also through land-based oil spills both on the North Slope and along the Trans-Alaska pipeline route. These spills have resulted from accidents, pipeline corrosion, and sabotage (Alyeska 2009; McBeath et al. 2008). Nevertheless, land-based oil spills from the Trans-Alaska pipeline and the North Slope oil industry complex have proven to be considerably more manageable than the 1989 *Exxon Valdez* marine oil spill disaster examined in chapter 4. The technological and logistical problems of oil spill response are far more daunting at sea than they are on land. In that sense, the marine oil transportation component of the Trans-Alaska Pipeline System has so far proven to be the weakest link in the chain that transports oil from northern Alaska to markets outside Alaska.

The federal and state right-of-way agreements for the Trans-Alaska pipeline anticipated that there would come a time when the North Slope oil fields would no longer produce sufficient oil to support continued operation of the pipeline. These right-of-way agreements required that the Trans-Alaska pipeline be dismantled (and that the lands in the pipeline route be ecologically restored) once the pipeline ceased operations (Fineberg 2004; GAO 2002, 74). While this requirement for eventually removing the Alaska pipeline and restoring the lands in the pipeline route seems clear, concerns have been raised that much of the North Slope oil industry complex will not be dismantled and the land area it occupies will not be restored to its natural condition once oil production ceases in that region (National Research Council 2003).

A potential source of industry and political resistance to the full environmental remediation of the North Slope oil industry complex is the high cost of removing infrastructures and restoring ecosystems in the remote and sensitive environment of the North Slope. In the event of a largely unmitigated abandonment of the North Slope oil infrastructures, one of the largest industrial complexes in the world would be left in place as an industrial wasteland in Arctic Alaska. Natural recovery of disturbed habitats in Arctic Alaska is very slow due to the harsh conditions found in that region, so the possibility of largely unmitigated infrastructure abandonment on the North Slope holds the potential for centuries of harmful environmental effects in that region (GAO 2002; National Research Council 2003).

The effects of the Alaska pipeline punctuation were not limited to Alaska. The national and international impacts of the Alaska pipeline punctuation are examined next.

National and International Impacts of the Alaska Pipeline Punctuation

The Alaska pipeline punctuation authorized the building of oil infrastructures in Alaska that would become an important source of domestic oil production for the United States (Alaska Oil Spill Commission 1990). On the international level, events occurring in the early stages of the Alaska pipeline punctuation triggered a cascade of policy reforms in Canada that progressively expanded Canadian claims to maritime jurisdiction in the Arctic. The test voyages of the icebreaking tanker *Manhattan* included passages along the northern coastline of Canada (Griffiths 1987). These Arctic tanker tests demonstrated that the northern coastline of Canada could be exposed to oil spills and other risks to the environment associated with commercial shipping. The Arctic tanker tests also raised concerns in Canada over the issue of maritime jurisdiction in the Arctic region (Coen 2012; Grant 2010; Griffiths 1987). The northern Canadian coastline included an extensive archipelago of islands named the Canadian Arctic archipelago. Canada claimed jurisdiction over the waters surrounding the islands of the Canadian Arctic archipelago, but much of this claim was not recognized by the United States (Kirkey 1996, 42). The Arctic tanker tests caused a public and political outcry in Canada, as the tests were widely perceived as an affront to Canadian claims of maritime jurisdiction in the Arctic (Elliot-Meisel 2009; Griffiths 1987). In 1970, the Canadian government responded to this outcry by proposing reforms designed to expand Canada's maritime jurisdictional claims. These reforms included the Arctic Waters Pollution Prevention Act, which asserted the right of Canada to enforce antipollution regulations within a zone extending 100 nautical miles seaward from the northern Canadian coastline (Elliot-Meisel 2009, 211; Grant 2010; Griffiths 1987, 90–91; Kirkey 1996, 49). A nautical mile is a standard measure of distance at sea; 1 nautical mile is equivalent to 1.15 land or statute miles (National Institute of Standards and Technology 2011; Sohn et al. 2010, 4). Antipollution regulations subsequently developed under the Arctic Waters Pollution Prevention Act included requirements for vessel design and navigational procedures, and also included the power to halt vessels and seize their cargo (Grant 2010, 356).

Canada subsequently attempted to reinforce its claims to Arctic maritime jurisdiction through further unilateral actions and international negotiations. The Canadian government negotiated a provision in the 1982 United Nations Convention on the Law of the Sea that secured widespread international recognition for the Arctic waters pollution prevention zone previously claimed by Canada (Griffiths 1987). The 1982 Law of the Sea Convention included a provision allowing a nation to extend its jurisdiction over the pollution control of vessels in ice-covered waters throughout an exclusive economic zone that generally extended out to two hundred nautical miles from its coastline (Griffiths 1987; Sohn et al. 2010). The Law of the Sea Convention therefore not only gave widespread international recognition to the Arctic waters pollution prevention zone previously claimed by Canada but also created a basis for Canada to further extend that zone. The United States did not join the Law of the Sea Convention, and a bilateral disagreement over the status of the waters of the Canadian Arctic archipelago has endured for decades following the Arctic voyages of the *Manhattan* (Coen 2012).

In 1985, the United States Coast Guard vessel *Polar Sea* completed a voyage along the northern coastline of Canada, setting off another public and political outcry in Canada concerning maritime jurisdiction in the Arctic (Elliot-Meisel 2009, 212). The Canadian government responded in 1986 by drawing a boundary line around the outer perimeter of the Canadian Arctic archipelago and declaring the waters within that boundary line as internal waters of Canada where the Canadian government claimed full sovereignty, including the authority to deny transit access to foreign vessels (Coen 2012; Grant 2010, 213, 377–79; Griffiths 1987).

The expanding Canadian claims to maritime jurisdiction in the Arctic have long been at odds with the U.S. position favoring freedom of navigation in the Arctic region of North America, a position that serves the security and commercial interests of the United States insofar as it allows freedom of navigation for U.S. vessels in the Arctic. This position also serves the interests of the United States by discouraging a precedent for further regulation or blockage of international straits by the nations bordering those straits—a precedent perceived by the U.S. government as a major threat to international trade and naval maneuvers (Elliot-Meisel 2009; Grant 2010; Griffiths 1987).

Canada's 1970 Arctic maritime jurisdiction initiatives reinforced U.S. arguments for a North Slope oil route that avoided Canadian involvement. The regulatory and political implications of the Arctic tanker route were greatly complicated by the Canadian claim of regulatory authority over tankers sailing along the Arctic coastline of Canada (Coen 2012; Griffiths 1987). The 1970 initiatives to expand Canadian maritime jurisdiction in the Arctic also demonstrated that Canada was willing to take swift unilateral actions for environmental protection despite strong objections from the United States on matters of joint commercial and strategic interest. The 1970 Canadian Arctic maritime jurisdiction initiatives thereby signaled that Canada might pursue unilateral actions for the environmental management of a Trans-Canada pipeline against U.S. strategic and commercial interests, a possibility of great importance because much of the Trans-Canada route would be under Canadian jurisdiction. In sum, the Canadian claims to jurisdiction in Arctic waters reinforced arguments against both the Arctic tanker route and the Trans-Canada route for North Slope oil (Grant 2010).

Summary

The Alaska pipeline punctuation began with the announcement of a major oil field discovery on the North Slope of Alaska in 1968. This critical period of reform continued through the approval of the Alaska Native Claims Settlement Act in 1971 (clearing the route for the Alaska pipeline) and ended with the approval of the Trans-Alaska Pipeline Authorization Act in 1973 (authorizing the building of the Alaska pipeline). These two federal laws together incorporated a multifaceted policy image that supported the building of the Alaska pipeline, the settlement of Alaska Native land claims, and land conservation planning across vast areas in Alaska. These two federal laws are the institutional arrangements that enabled the construction of a major oil transportation system in Alaska, and these enduring institutional arrangements have led to enduring policy consequences; for more than three decades, the Trans-Alaska Pipeline System has simultaneously served as a major economic asset for Alaska and as a major environmental threat to the areas of Alaska along its route.

The institutional arrangements and policy image established in this critical period of reform contained the elements of a compromise that distributed benefits among development interests, Alaska Native interests, and environmental interests. The building of the Trans-Alaska Pipeline System would provide unprecedented economic benefits to Alaska, a major resource for the oil industry, and an important supply of oil for the nation. The Alaska Native Claims Settlement Act would provide a major land title gain for the Alaska Natives, yet also create the basis for large-scale land conservation planning in Alaska that would serve the interests of the environmental movement. The critical period of policy reform that authorized the Trans-Alaska Pipeline System therefore created institutional arrangements that broadly distributed the benefits of the natural resources of Alaska. However, this critical period of reform also established institutional legacies with long-enduring consequences for environmental protection in Alaska. The operations of the North Slope oil industry complex and the Trans-Alaska Pipeline System have long posed major hazards to the environment in Alaska.

The critical period of reform that allowed the building of the Alaska pipeline was a significant defeat for the U.S. environmental movement, as the environmentalists failed to either block the Alaska pipeline or route the pipeline through Canada to avoid the use of oil tankers in the system. But this critical period was not an unmitigated defeat for the American environmental movement. First, the critical period established a series of federal requirements for the design and construction of the Alaska pipeline that reduced the risk of oil spills from the pipeline (Coates 1993, 231; Siehl and Want 1972; Willis 2010). Second, this critical period planted the seeds of future victories for the environmental groups promoting land conservation in Alaska by establishing a systematic process of conservation planning in Alaska. The proposals for conservation planning in Alaska during the debate over the Alaska Native land claims marked the beginning of a prolonged critical period of reform concerning nature conservation in Alaska, which is examined in chapter 3.

The Alaska National Interest
Lands Conservation Act

T HE ALASKA NATIVE CLAIMS SETTLEMENT ACT played important roles in both the Alaska pipeline punctuation and in a subsequent punctuation concerning Alaska lands conservation. A 1978 House of Representatives report noted that the Alaska Native Claims Settlement Act in 1971 "set in motion a sequence of events which may well constitute the most significant single land conservation action in the history of our country" (H.R. Rep. No. 95-1045 Part I 1978, 70). Following the enactment of the Alaska Native Claims Settlement Act, Interior Secretary Rogers Morton implemented the land conservation planning sections of that law by withdrawing vast areas of Alaska for conservation planning purposes, thereby blocking claims on those areas by the state of Alaska (Allin 2008; Nelson 2004; Ross 2000). The Alaska Native Claims Settlement Act included two provisions that together established a process for the Interior secretary to withdraw federal lands in Alaska for conservation planning purposes. Section 17(d)(1) of the Alaska Native Claims Settlement Act allowed the withdrawal of federal lands in Alaska with no discernible limits on either the acreage to be withdrawn or the amount of time for which that acreage could be withdrawn. The broad authority granted by section 17(d)(1) was confusingly combined with the more limited authority granted by section 17(d)(2), which specified that no more than 80 million acres could be withdrawn in

Alaska for a period of no more than five years by the Interior secretary (Allin 2008, 217–18; Williss 2005).

The implementation of the conservation planning sections of the Alaska Native Claims Settlement Act involved input from federal land management agencies, environmentalists, the state of Alaska, and a joint federal-state land use planning commission established by the Act. The National Park Service had previously studied areas of conservation interest in Alaska (Cahn 1982; Williss 2005). Following the enactment of the Alaska Native Claims Settlement Act, additional land use planning studies of Alaska were conducted by federal agencies (including the National Park Service, Fish and Wildlife Service, and Forest Service) to support agency recommendations for Alaska land withdrawals and legislative proposals for Alaska land management (S. Rep. No. 96-413 1979, 131–32). For example, the National Park Service produced plans and environmental impact statements to support its proposals for new national park units in Alaska (Williss 2005, 60–63). The National Park Service and the Forest Service produced competing plans for federal land management in Alaska, recapitulating a long history of competition between these two services for jurisdiction over various federal lands across the nation (Hays 2007; Miles 2009; Williss 2005). In addition to these agency studies, the joint federal-state land use planning commission provided its own separate studies and recommendations concerning land withdrawals in Alaska to the Interior secretary. The Alaska Native Claims Settlement Act therefore set in motion a multifaceted government effort for land use planning in Alaska. As a result, the Alaska lands conservation debate was informed by a current and extensive collection of government studies of land use planning in Alaska (S. Rep. No. 96-413 1979, 131–33; The Wilderness Society 2001, 17; Williss 2005, 134). The process of Alaska lands conservation planning also involved contributions from environmental groups that collaborated closely with state and federal agencies. In 1969, representatives of conservation groups, federal agencies, and state agencies held a wilderness workshop in Alaska. This wilderness workshop led to the formation of a group named the Alaska Wilderness Council that acted as a forum through which environmentalists, federal officials, and state officials collaborated to develop proposals for land conservation in Alaska (Cahn 1982, 10–14; Ross 2000, 194; Williss 2005). As a result, the Alaska lands conservation planning process was informed by proposals developed with input from the environmental movement.

The Alaska Native Claims Settlement Act lifted the freeze on state land selections in Alaska for ninety days, and during that brief period the state of Alaska rushed to select approximately 77 million additional acres in Alaska as the process of federal land withdrawals was under way (Allin 2008, 218–19;

Williss 2005, 53). These additional state land selections overlapped in part with areas Interior Secretary Morton intended to withdraw. The state of Alaska pursued litigation to uphold its selections; the case was settled in 1972 through a memorandum of understanding between the state of Alaska and the federal government under which the state was allowed to select 42 million additional acres in Alaska (H.R. Rep. No. 95-1045 Part I 1978, 346–47, 403–11; S. Rep. No. 95-1300 1978, 389–90; Williss 2005). In 1972, Interior Secretary Morton announced final land withdrawals in Alaska under sections 17(d)(1) and 17(d)(2) of the Alaska Native Claims Settlement Act. First, 79 million acres of Alaska were withdrawn under section 17(d)(2) for a period of five years. Second, both the lands withdrawn under section 17(d)(2) and additional lands in Alaska were withdrawn under section 17(d)(1) for an indefinite period (Allin 2008, 218, 237; H.R. Rep. No. 95-1045 Part I 1978, 193–94; Williss 2005, 59, 102). In 1973, Secretary Morton announced legislative proposals for 83 million acres of new federal conservation units in Alaska, with the majority of this conservation acreage to be added to the National Park and National Wildlife Refuge Systems (Cahn 1982, 14; Williss 2005, 67–68).

Following the announcement of legislative proposals for Alaska lands conservation by Secretary Morton, legislation for that purpose was introduced in Congress but did not advance into law. Legislation for Alaska lands conservation was not a priority for either the Nixon or Ford administrations (Allin 2008, 219–20; Cahn 1982; Swem and Cahn 1984; Williss 2005). By contrast, Alaska lands conservation became the top environmental priority of the administration of President Jimmy Carter. The cause of Alaska lands conservation would also be given a well-positioned champion in Congress when Representative Morris Udall became chairman of the House of Representatives Committee on Interior and Insular Affairs (Allin 2008, 221–22). Nonetheless, the development of Alaska lands conservation legislation in Congress would lead to a prolonged political conflict involving a number of different bills on this subject. Major issues in the ensuing Alaska lands conservation debate included the total area to be conserved, the location and character of the lands to be conserved, the allowable uses of conserved areas, and the management systems and agencies that would provide the administrative framework for conservation in Alaska. As described later in this chapter, the complexity of the debate reflected in part the diverse purposes and histories of different federal land management systems and agencies that might be used for conservation purposes in Alaska (Vale 2005).

The debate over Alaska lands conservation was heavily influenced by the Alaska Coalition of environmental organizations that had first formed in 1971

(Allin 2008, 221; Mitchell 2001, 444; Williss 2005, 36). The Alaska Coalition aimed to promote major land conservation actions in Alaska, drawing on the experience gained from past successes of the U.S. environmental movement (Duscha 1981, 8). The membership of the Alaska Coalition had grown over time. During the Alaska lands conservation debate the Alaska Coalition included a large number of prominent environmental groups such as The Wilderness Society, Sierra Club, Friends of the Earth, National Audubon Society, Natural Resources Defense Council, Environmental Defense Fund, National Parks and Conservation Association, National Wildlife Refuge Association, and the World Wildlife Fund (Cahn 1982). The Alaska Coalition succeeded in creating a highly organized national campaign to promote Alaska lands conservation (Allin 2008, 224; Duscha 1981; Nash 2001; Turner 2007). However, proposals for large-scale land conservation in Alaska met with sustained opposition from the Alaska congressional delegation and development interests (Carson and Johnson 2001; Swem and Cahn 1984). The result was a prolonged conflict in Congress between the cause of Alaska conservation and the cause of Alaska development, a conflict that is examined next by comparing and contrasting two competing bills addressing Alaska lands that emerged during the critical period. These bills provide insight into the initial bargaining positions and interests of those favoring conservation and those favoring development in Alaska (Allin 2008; Naske and Slotnick 2011).

The Cause of Alaska Lands Conservation in Congress

Environmentalists from the Alaska Coalition worked with staff and elected officials of the House of Representatives Committee on Interior and Insular Affairs to draft a major Alaska lands conservation bill (H.R. 39) introduced by Representative Udall in 1977 (Allin 2008, 220; Cahn 1982, 16; Williss 2005, 88). This conservationist bill was an ambitious plan for large-scale nature preservation in Alaska that expanded on the legislative proposals for Alaska lands conservation previously submitted by Interior Secretary Morton. The original H.R. 39 bill also contained provisions meant to authorize subsistence activities and sport hunting in some of the conservation areas it aimed to establish in Alaska (Cahn 1982, 15–16; Norris 2002; Williss 2005, 88). Government studies and congressional hearings had revealed that subsistence was essential to the way of life and cultures of Alaska Natives and other residents of rural areas of Alaska (H.R. Rep. No. 95-1045 Part I 1978, 181–83). A 1978 Senate report noted that "Alaska's more than 200 rural villages are unique in that they are the last communities in the United States in which a substantial number of residents are still dependent upon the harvest of renewable resources on the public lands for their sustenance" (S. Rep. No. 95-1300 1978, 195). Subsistence activities had been permitted in several wildlife ranges and refuges administered by the Fish and Wildlife Service in Alaska prior to the Alaska lands debate, creating a precedent for the establishment of other federal conservation units that accommodated subsistence in Alaska (Federal Field Committee for Development Planning in Alaska 1968, 446). Furthermore, environmental groups had collaborated with representatives of Alaska Native groups in developing the H.R. 39 legislative proposal. The conservationist bill also aimed to create conservation areas where sport hunting could continue, as sport hunting was a major recreational activity in Alaska (Cahn 1982, 15–16; Norris 2002; The Wilderness Society 2001).

The development of H.R. 39 was informed not only by extensive government studies and the strongly expressed conservation interests of the national environmental movement, but also by wide-ranging public hearings held across Alaska and at other sites across the nation. Members of the House of Representatives Committee on Interior and Insular Affairs conducted an extensive review of the Alaska lands conservation issue, including public hearings held by a subcommittee on General Oversight and Alaska Lands chaired by Representative John Seiberling of Ohio (Nelson 2009, 126). More than two thousand people gave oral testimony as part of these public hearings (H.R. Rep. No. 95-1045 Part I 1978, 74–77). Public hearings at five sites outside of Alaska revealed overwhelming support for large-scale Alaska lands conservation among those submitting testimony; by comparison, an extensive series of public hearings throughout Alaska revealed a more even balance of support for and opposition to large-scale Alaska lands conservation among those submitting testimony (H.R. Rep. No. 96-97 Part I 1979, 392–93; U.S. Congressional Record 1978, 14148–49).

The original H.R. 39 bill reflected a proposed policy image focused on creating a highly protected system of conservation units across very large scales in Alaska. H.R. 39 proposed the addition of 110.8 million acres to the National Park System and National Wildlife Refuge System in Alaska, as well as the addition of 4.1 million acres to the National Wild and Scenic Rivers System in Alaska. H.R. 39 also proposed wilderness designations across 146.5 million acres in both new and preexisting conservation units in Alaska. Most forms of development would be precluded in the areas designated as wilderness. These wilderness designations would be a particularly radical reform of land use policies in Alaska, as wilderness areas protected by statute encompassed only an estimated 75,000 acres of Alaska in 1978 (Allin 2008, 228; S.

Rep. No. 95-1300 1978, 414). While H.R. 39 would have diminished natural resource development opportunities in Alaska, it also held some potential economic benefits for Alaska. New conservation units in Alaska held the potential to become major destinations for tourists and thereby aid the tourism industry in Alaska, which was growing rapidly during the Alaska lands conservation punctuation (S. Rep. No. 96-413 1979, 130). A 1978 House of Representatives report noted that new conservation units in Alaska "should offer a tremendous boost to the recreation industry in Alaska" (H.R. Rep. No. 95-1045 Part I 1978, 71).

The policy image of the conservationist H.R. 39 bill focused on establishing wilderness as a leading land use in Alaska and on greatly expanding the areas under the jurisdictions of the National Park Service and the Fish and Wildlife Service in Alaska. The conservationist policy image thereby emphasized the use of institutional arrangements for land management that favored conservation over most forms of development in Alaska conservation units. First, the conservationist bill aimed to greatly expand the size of the national parks in Alaska under the management of the National Park Service. The National Park Service Act of 1916 had authorized the creation of the National Park Service as a dedicated management agency for the preexisting national park system (Runte 2010, 95). The first national park to be established after the founding of the National Park Service was Mount McKinley National Park, which was created by Congress in 1917 as the first large national park in Alaska (Norris et al. 1999, 44–45). The dual mission of the National Park Service was to preserve the parks for future generations, while also providing for the public enjoyment of the parks (Runte 2010, 95–96). The purposes of preservation and tourism conflicted at times, as the National Park Service built roads and visitor facilities in many parks with the aim of bringing large numbers of visitors into those parks. Tourism and the construction of tourism-related infrastructures caused some environmental damage in the national parks, but many other forms of development (such as mining and logging) were generally excluded from the national parks (Cahn 1982; Runte 2010; Swem and Cahn 1984). Overall, the National Park Service had a history of placing an exceptionally high priority on environmental protection in the lands under its jurisdiction when compared to the other federal land management agencies discussed in this chapter (Cahn 1982; Swem and Cahn 1984). By the time of the Alaska lands debate, the National Park Service administered 7.5 million acres of Alaska (H.R. Rep. No. 95-1045 Part I 1978, 69; Williss 2005). The National Park Service had actively pursued both planning and lobbying for the creation of large new national parks in Alaska that would protect the extraordinary wilderness and scenic values of Alaska on large scales (Miles 2009; Runte 2010).

The original H.R. 39 also aimed to greatly expand the size of the national wildlife refuge system in Alaska under the management of the Fish and Wildlife Service. The federal government had a long history of protecting wildlife habitat in Alaska (Cahn 1982). A 1978 Senate report noted that "the conservation of wildlife by federal withdrawal of their habitats under protective management began in Alaska in 1869 when the Congress established the Pribilof Reservation for the protection of the northern fur seal" (S. Rep. No. 95-1300 1978, 146–47). This early example of a wildlife reservation on the Pribilof Islands of Alaska would later be followed by the development of a coherent national system of wildlife refuges with a dedicated management agency. A national system of wildlife reservations on federal lands was first established in the early twentieth century by President Theodore Roosevelt through a series of executive orders. Roosevelt had begun this series of conservation initiatives by inquiring about the legal authority available for creating a federal bird reservation on Pelican Island off the coast of Florida. When informed that nothing in the law either authorized or prevented such an action, Roosevelt seized the initiative and signed an executive order in 1903 that established a federal bird reservation on Pelican Island. Pelican Island would be the first of many such reserves that formed the initial core of a national wildlife reservation system on federal lands; in all, Roosevelt established more than fifty federal wildlife reserves by executive actions in the years 1903–9, including several such reserves in Alaska (Brinkley 2009, 13–19; Brinkley 2011, 62–64; Fischman 2003, 35).

In 1940, a single agency to manage the national wildlife refuge system was established through the executive merger of the Bureau of Biological Survey and the Bureau of Fisheries to create the Fish and Wildlife Service (Fischman 2003, 39; Norris et al. 1999). The National Wildlife Refuge System Administration Act of 1966 established a consolidated statutory basis for the National Wildlife Refuge System (Fischman 2003, 45–53, 228–33). The core mission of this wildlife refuge system was to protect wildlife and habitat in the refuges, but wildlife refuges were nonetheless open to a greater range of uses than national parks. Hunting was generally banned in national parks but generally allowed in wildlife refuges, and hunters formed the leading constituency of the Fish and Wildlife Service (Cahn 1982, 12–13; Kaye 2006, 129). Natural resource extraction activities were also permitted to a greater extent in the wildlife refuges than in national parks (Cahn 1982, 12–13; Fink 1994; Fischman 2003, 2005; Layzer 2011, 114). For example, half of the Kenai National Moose Range of Alaska had been opened to oil exploration in 1958 (Coates

1993, 91–95). A statement by five Senators in a 1978 Senate report noted that "National Wildlife Refuges in Alaska today have grazing, oil and gas leases, and timber sales in some areas" (S. Rep. No. 95-1300 1978, 402).

The Fish and Wildlife Service administered 20 million acres of Alaska at the time of the Alaska lands debate (H.R. Rep. No. 95-1045 Part I 1978, 69). A number of national wildlife conservation units had been established across Alaska through executive orders issued by presidents and public land orders issued by Interior secretaries (Woods 2003). However, a wildlife conservation unit established by executive action could also be diminished or revoked by another executive action (Woods 2003). Conservationists therefore sought a land bill that would both expand the national wildlife refuge system of Alaska and provide durable statutory protection for the system. The vast scale, wild character, and abundant wildlife of Alaska created a unique opportunity for the expansion of the national wildlife refuge system to encompass exceptionally large and relatively undisturbed wildlife habitats. The enlargement of the national wildlife refuge system in Alaska could also offer added protection for a number of species that were found in Alaska but nowhere else in the nation, for species that migrated to and from Alaska to other points in the nation, and for species that migrated to and from Alaska across international borders and that were protected by treaties between the United States and other nations (S. Rep. No. 95-1300 1978, 146–48, 168–69, 398–400).

The most ambitious goal of the conservationist bill was the creation of vast new wilderness designations across new and preexisting conservation units in Alaska, as the wilderness designation was the highest level of protection available for federal lands. The national wilderness system was the culmination of a long campaign by the American environmental movement to create a system of conservation units that would exclude the development of roads and other infrastructures. The long political campaign to establish a wilderness system had its origins in the founding of the environmental organization The Wilderness Society in 1935 (Sutter 2002, 6–7). A central concern of the founders of The Wilderness Society was the impact of roads and automobiles on the wilderness areas of the nation. The rise of the road in American society was clearly a challenge to wilderness values, yet even the National Park Service invested heavily in road building in the national parks in the interest of bringing more visitors into the parks (Sutter 2002). The Wilderness Society joined with other conservation organizations to promote legislation for a new system of lands where roads and most other forms of development would be prohibited. These efforts eventually led to a federal statute establishing a national wilderness system (Allin 2008; Harvey 2005; Miles 2009; Nash 2001).

The Wilderness Act of 1964 (Pub. L. 88-577) established the National Wilderness Preservation System and authorized Congress to designate federal lands as wilderness areas that would be protected and managed to preserve their natural conditions. The wilderness designation was the highest form of land protection available under federal law. Roads, structures, installations, commercial enterprises, and mechanized transport were generally not permitted in wilderness areas (although the Wilderness Act incorporated some limited exceptions to this general ban on mechanized transport and commercial activities in wilderness areas). As a category of federal land protection that could be superimposed upon the other federal land management systems, wilderness areas could be administered by any of the four major federal land management agencies: the National Park Service, Fish and Wildlife Service, Forest Service, and Bureau of Land Management (Allin 2008; Miles 2009; Nash 2001). The Alaska lands debate gave the American environmental movement its greatest single opportunity to expand the National Wilderness Preservation System.

The H.R. 39 bill also sought to create a large new system of wild and scenic rivers in Alaska. The National Wild and Scenic Rivers System had been previously established by the Wild and Scenic Rivers Act of 1968 (Pub. L. 90-542). The purpose of the National Wild and Scenic Rivers System was to maintain some areas of rivers in a free-flowing and wild state, and to protect some areas of riverbanks as well. A particular concern addressed by the Wild and Scenic Rivers Act was the environmental damage caused by dams. The Wild and Scenic Rivers Act stated that "the established national policy of dam and other construction at appropriate sections of the rivers of the United States needs to be complemented by a policy that would preserve other selected rivers or sections thereof in their free-flowing condition to protect the water quality of such rivers and to fulfill other vital national conservation purposes" (82 Stat. 906). In a policy design that paralleled the National Wilderness Preservation System, the National Wild and Scenic Rivers System was superimposed upon the other federal land management systems and could therefore be administered by all four of the major federal land management agencies.

The environmental threats posed by dams to rivers and their surrounding areas were exemplified by a proposal for a major dam project in Alaska that received considerable attention from the federal government prior to the Alaska lands conservation punctuation (Coates 1993; Ross 2000; Willis 2010). The proposed Rampart Canyon dam on the Yukon River of Alaska would have created a reservoir covering 6.7 million acres in Alaska, and in so doing would have destroyed large areas of wildlife habitat (U.S. Fish and

Wildlife Service 1964). The federal government withdrew a large area around the Yukon River for the proposed Rampart dam reservoir (Federal Field Committee for Development Planning in Alaska 1968). The Rampart dam reservoir would have been located in the same general area as the Yukon Flats National Wildlife Refuge that was established in 1980 (see map 3.1 for the location of this refuge). The Rampart dam proposal was met with widespread opposition. The U.S. Bureau of Reclamation opposed the Rampart dam project, and sought instead to build a smaller dam elsewhere in Alaska. National environmental groups and outdoor sporting groups opposed the Rampart dam project due to the massive loss of wildlife habitat that the Rampart reservoir threatened to cause. Canada opposed the Rampart dam project due to concerns over damage to wildlife and navigation that the project threatened to cause. Members of seven Alaska Native villages that would have been submerged by the Rampart reservoir opposed the Rampart dam project as well (Coates 1993; Ross 2000; Willis 2010). The Fish and Wildlife Service objected vigorously to the Rampart dam proposal, stating in a 1964 report on the Rampart dam and reservoir project that "nowhere in the history of water development in North America have the fish and wildlife losses anticipated to result from a single project been so overwhelming" (U.S. Fish and Wildlife Service 1964, 8). The Rampart dam proposal lost support in the federal government and was effectively abandoned in 1971 (Coates 1993, 154). However, the Rampart dam proposal clearly demonstrated that rivers and their surrounding areas could be threatened by development in Alaska. The Alaska lands conservation punctuation created an opportunity to establish protections for rivers and their surrounding areas across large scales in Alaska.

In sum, the H.R. 39 bill reflected a proposed policy image focused on conserving wilderness and wildlife values across very large scales in Alaska. This conservationist position found substantial support in Congress. A 1979 House of Representatives report referred to wilderness as "the singular distinguishing characteristic of the public lands of Alaska" (H.R. Rep. No. 96-97 Part I 1979, 473) and further noted that "one of the greatest values of America's heritage in Alaska is the wilderness character of the land" (H.R. Rep. No. 96-97 Part I 1979, 474). During the congressional debates concerning the Alaska lands issue, Representative Udall stated that "wilderness is what these lands in Alaska are all about" (U.S. Congressional Record 1979, 46), and he went on to note the exceptional opportunity available for large-scale ecosystem protection on land in Alaska: "It speaks to our responsibility as stewards of the land, to provide future generations not merely fragmented remnants of our natural heritage, but with whole, intact, truly magnificent ecosystems . . .

Not in our generation, nor ever again, will we have a land and wildlife conservation opportunity approaching the scope and importance of this one" (*U.S. Congressional Record* 1979, 48).

In contrast to the conservationist policy image embedded in H.R. 39, counterproposals by the Alaska congressional delegation reflected a proposed policy image that gave far less weight to conservation and far greater weight to development opportunities in Alaska. The essence of the objections to H.R. 39 by elected officials from Alaska was their interest in maintaining broad opportunities for natural resource extraction and other forms of commercial development on Alaska lands. In practice, this meant that Alaska lands legislation proposed by the Alaska congressional delegation would designate far less land for conservation purposes than Alaska lands legislation supported by the Alaska Coalition. Alaska lands legislation proposed by the Alaska congressional delegation also contained land management provisions far more favorable to commercial development than Alaska lands legislation supported by the Alaska Coalition. This development-oriented perspective is considered next through an examination of an Alaska lands bill proposed by two members of the Alaska congressional delegation.

The Cause of Alaska Development in Congress

The proposed policy image for Alaska lands legislation promoted by the Alaska congressional delegation included a relatively limited expansion of federal land protections in Alaska when compared to the conservation provisions of the original version of H.R. 39. A bill supported by Senator Ted Stevens and Representative Don Young of the Alaska delegation proposed the addition of 25.2 million acres to conservation systems in Alaska (only 22 percent of the area proposed for addition to conservation systems in Alaska by the original H.R. 39). The conservation additions in Alaska proposed by the Stevens-Young bill included 18.5 million acres to be added to the National Park System and National Wildlife Refuge System, 5.7 million acres to be added to the National Forest System, and 1 million acres to be added to the National Wild and Scenic Rivers System. Most importantly, this Stevens–Young bill would have added no wilderness designations at all in Alaska (a stark contrast with the 146.5 million acres of wilderness designations across new and preexisting conservation units in Alaska proposed in the original H.R. 39). The Stevens–Young bill therefore gave much less support to conservation and much more support to development than did H.R. 39. Given the absence of new wilderness designations in the Stevens–Young bill, the history of federal land management suggested that natural resource development would be allowed across most federal lands in Alaska under this bill (Allin 2008, 228).

The national forest system provided a clear example of the type of federal land management that favored development and that was therefore favored by the Alaska congressional delegation and their allies in Congress. The national forest system had its origins in congressional and presidential actions to reserve forests on federal lands to prevent uncontrolled timber cutting on those lands. Nineteenth-century federal land laws had often promoted a policy of land disposal by which large areas of federal lands were transferred into private ownership. A significant reform to this policy of federal land disposal occurred in the General Land Law Revision Act of 1891, which included a provision generally referred to as the Forest Reserve Act that empowered the president to proclaim forest reserves on federal lands. The Forest Reserve Act established executive authority to protect federal forests from the rampant timber cutting that was causing large-scale deforestation in America, but the law did not provide for the management of these forest reserves. Federal authority to actively manage these forest reserves was subsequently established in the Forest Management Act of 1897 (Andrews 2006, 97–106; Hays 1999, 36, 44).

The national forest system was built up through forest reservations made by several presidents under the Forest Reserve Act. The greatest additions to the forest reserve system were made by President Theodore Roosevelt, including two large additions in Alaska (Hays 1999, 47). President Roosevelt established the Tongass National Forest and the Chugach National Forest in Alaska in the period 1902–8 (Brinkley 2011, 53–58; H.R. Rep. No. 95-1045 Part I 1978, 157–58; Ross 2000, 230). In 1905, the Forest Transfer Act transferred the forest reserves from the U.S. Department of the Interior to the U.S. Department of Agriculture (USDA). The USDA Forest Service was given the task of managing the forest reserves, which were renamed national forests (Andrews 2006, 144–45). In 1907, Congress enacted restrictions on the power of the president to proclaim additional forest reserves, but by the time these new restrictions became law, the combined effect of presidential forest reserve proclamations and the 1905 transfer of forest reserves to the USDA had established a large national forest system under the jurisdiction of the Forest Service (Hays 1999, 47).

By the time of the Alaska lands conservation punctuation, the Forest Service administered 20.7 million acres of Alaska (H.R. Rep. No. 95-1045 Part I 1978, 69). The Forest Service had by that time accumulated a lengthy history of managing the national forests with an emphasis on the extractive use of natural resources, including timber cutting and grazing in the national forests (Hays 2009). The Forest Service had built a vast network of roads in the national forests to promote private timber cutting, and the combination of road building and timber cutting activities caused large-scale environmental damage in the national forest system (Andrews 2006, 194; Davis 2001; Layzer 2011). The overall effect of Forest Service policies for road building and timber sales was to provide public subsidies for the private use of natural resources in the national forests. The Forest Service's emphasis on natural resource extraction was exemplified by heavily subsidized timber cutting in the Tongass National Forest of Southeast Alaska (H.R. Rep. No. 95-1045 Part I 1978, 158–59). A statement by five members of Congress in a 1978 Senate report noted that "the Forest Service's record in Southeast Alaska and the type of management scenarios they have suggested for interior Alaska areas indicate that commodity development in new national forests would be heavily subsidized and promoted at the expense of fish and wildlife values" (S. Rep. No. 95-1300 1978, 402). Environmental interests generally opposed the natural resource extraction policies of the Forest Service, and the original H.R. 39 bill would not have added to the National Forest System in Alaska (Allin 2008; Hays 2007, 2009).

The Stevens–Young bill would have left large areas of Alaska that environmentalists aimed to conserve under the management of the state of Alaska and the Bureau of Land Management. The state of Alaska aimed to develop some of those lands, and the history of the Bureau of Land Management also evinced considerable agency support for development. The Bureau of Land Management was formed in 1946 through the executive merger of the General Land Office and the Grazing Service (Skillen 2009). At the time of the Alaska lands conservation debate, the Bureau of Land Management administered most of the area of federal lands found in Alaska (H.R. Rep. No. 95-1045 Part I 1978, 69). The Bureau of Land Management pursued a multiple-use policy that allowed a wide range of natural resource development activities (including logging, grazing, mining, and fossil fuel extraction) on federal lands (S. Rep. No. 96-413 1979, 388). The Bureau of Land Management also administered conservation areas, but the capacity of the bureau to adequately conserve land and wildlife was already in question during the period of the Alaska lands debate. The concept of National Conservation Areas administered by the Bureau of Land Management had been formally established in 1970, but these areas allowed for multiple uses beyond conservation (H.R. Rep. No. 96-97 Part I 1979, 259). A 1976 Senate report noted substantial evidence of deteriorating ecological conditions on most of the area of public grazing lands administered by the Bureau of Land Management across the nation (S. Rep. No. 94-593 1976). This Senate report also called into question the ability of the Bureau of Land Management (BLM) to adequately manage wildlife:

> Clearly, the record of BLM's wildlife management has not been an encouraging one. The reason for this undoubtedly arises from the fact that BLM has a number of other important missions such as mining, logging, livestock grazing, and fossil fuel development which often conflict with wildlife management. In performing these conflicting missions, BLM is unable to devote sufficient attention to the needs of wildlife. In short, its mission is not wildlife protection or enhancement.
>
> In contrast to BLM, the U.S. Fish and Wildlife Service has as its basic mission the protection and enhancement of wildlife (S. Rep. No. 94-593 1976, 4–5).

This negative assessment of the wildlife management capacities of the BLM was quoted in both House and Senate reports concerning the Alaska lands bill in 1979 (H.R. Rep. No. 96-97 Part I 1979, 422–23; S. Rep. No. 96-413 1979, 388).

In sum, the original H.R. 39 bill and the Stevens–Young bill were far apart in the scale and intent of their proposed federal land actions in Alaska. These two conflicting bills represented the opposite ends of the spectrum of political conflict among the numerous bills concerning Alaska lands introduced in Congress during the Alaska lands conservation punctuation. Environmental interests typically favored bills that emphasized the restriction of development across large scales in Alaska for conservation purposes. In contrast, development interests and elected officials representing Alaska typically favored bills emphasizing multiple uses of Alaska lands and providing limited restrictions on development in Alaska (H.R. Rep. No. 96-97 Part I 1979, 261; Swem and Cahn 1984). The original H.R. 39 bill proposed large-scale additions to the federal land conservation system in Alaska, emphasized land management by the conservation-oriented National Park Service and Fish and Wildlife Service, and proposed sweeping wilderness designations that would prevent industrial development across large areas of Alaska. By contrast to H.R. 39, the Stevens–Young bill proposed adding far less area to national conservation units in Alaska, promoted land management by the Forest Service and Bureau of Land Management with their histories of promoting and subsidizing natural resource development on federal lands under their jurisdictions, and proposed no wilderness designations in Alaska at all (S. Rep. No. 96-413 1979, 387–88). The widely differing viewpoints of those favoring conservation and those favoring development in Alaska would contribute to a prolonged congressional debate over Alaska lands conservation legislation.

Congressional Debate and Delay

The prolonged congressional debate over Alaska lands legislation produced many variations on the fundamental and competing themes of conservation and development in Alaska. A number of Alaska lands conservation bills were debated in the House and Senate, including a bill sponsored by the Carter administration (S. Rep. No. 95-1300 1978; S. Rep. No. 96-413 1979). The development of Alaska lands conservation legislation was characterized by a series of compromises designed to balance environmental interests with oil, gas, timber, mineral, subsistence, and sport hunting interests in Alaska. These compromises were largely accomplished by redrawing conservation unit boundaries and by establishing special management policies for some of these conservation units. The interests of the mining industry were served by allowing the continuing development of existing mining claims within the new conservation units in Alaska. Oil, gas, and mining interests were accommodated by drawing conservation unit boundaries to avoid incorporating many of the major oil, gas, and mining prospects in Alaska into conservation areas. Longstanding environmental interests in the conservation of northern Alaska were served by more than doubling the size of a preexisting Arctic wildlife conservation unit and designating part of that unit as a wilderness, while the interests of the oil and gas industry in northern Alaska were served by designating the coastal plain of this Arctic conservation unit as an area to be explored for its oil and gas potential (the subsequent policy conflict over oil and wilderness in this Arctic National Wildlife Refuge is explored in chapter 5). Subsistence interests were served by allowing the continuation of subsistence activities in some of the new conservation units in Alaska, and sport hunting interests were served by establishing large conservation areas where sport hunting would be allowed. Environmental interests in Southeast Alaska were served through the designation of substantial new wilderness areas in the Tongass National Forest, while the interests of the timber industry in Southeast Alaska were served by establishing a new funding line to support the timber industry in the Tongass. Furthermore, the interests of the mining industry in the Tongass were served by setting aside certain areas of the Tongass for mining exploration and development. Finally, the general interest in preserving reasonable options for access in a state with few roads was served by allowing the continuation of mechanized access for traditional uses in the new conservation units of Alaska (H.R. Rep. No. 95-1045 Part I 1978, 67–68; H.R. Rep. No. 96-97 Part I 1979, 139–40, 227, 594; Ross 2000; Williss 2005). Therefore, the development of Alaska lands legislation reflected a complex balance among numerous competing interests.

In 1978, the House of Representatives passed a revised H.R. 39 by a large margin, proposing the addition of 123.9 million acres of federal conservation units in Alaska (Allin 2008, 228–30). The revised H.R. 39 passed by the House had the support of every major environmental organization in the United States but was opposed by the entire Alaska congressional delegation (Andrus and Freemuth 2006; *U.S. Congressional Record* 1978, 14164–69). The Alaska governor and an overwhelming majority of the Alaska state legislature also opposed H.R. 39 (*U.S. Congressional Record* 1978, 14141, 14164). These elected officials from Alaska were joined in their opposition to large-scale land conservation in Alaska by various development interests (including oil, timber, and mining interests) in a position to lose opportunities for natural resource development due to land conservation in Alaska (Cahn 1982, 19; Carson and Johnson 2001; Swem and Cahn 1984; Williss 2005). The interests opposing large-scale conservation in Alaska supported a proposed policy image focused on maintaining opportunities for economic development in Alaska. The three members of the Alaska congressional delegation were joined in their opposition to H.R. 39 by other members of Congress that supported development interests (H.R. Rep. No. 95-1045 Part I 1978, 367–74). A statement by two representatives in a 1979 House report noted the development interest in Alaska: "Alaska is more than a visual feast, more than a vast wildlife habitat—it is a treasure chest of mineral wealth—both fuel and non-fuel" (H.R. Rep. No. 96-97 Part I 1979, 685). The strategy of opposition to the conservationist bill included extensive negotiations aimed at reducing the scale of the lands to be dedicated for conservation in Alaska for the purpose of accommodating oil, gas, mining, and timber interests (H.R. Rep. No. 95-1045 Part I 1978, 401). Senator Stevens of Alaska was deeply involved in this strategy of opposition through negotiation, which would prove successful in reducing the scale of conservation in the final Alaska lands bill for the potential benefit of resource development industries and the resource economy of Alaska. By contrast, Senator Mike Gravel of Alaska adopted a disruptive strategy of opposition through aggressive stalling maneuvers, filibuster threats, and filibusters—all of which would eventually be overridden by Congress (Andrus and Freemuth 2006; Cahn 1982; Duscha 1981; Nelson 2004). In a statement inserted into a 1979 Senate report, Senator Gravel protested the Senate version of the Alaska lands bill: "The vast land 'locked up' by the bill (over 100 million acres), the inadequate provisions for future access across conservation system units created by the bill, and the imposition of a huge new federal bureaucracy will effectively freeze Alaska permanently in the economic state it is [in] now, with limited opportunities for growth or diversification" (S. Rep. No. 96-413 1979, 436).

Several Senate versions of the Alaska lands bill provided substantially less protection for lands in Alaska than the House versions of the bill. In a 1979 Senate report, Senators Howard Metzenbaum of Ohio and Paul Tsongas of Massachusetts protested the removal of a number of proposed wilderness areas from a Senate version of the Alaska lands bill: "Wilderness is the very essence of the parklands and wildlife refuges the American people want protected in Alaska. Here, as nowhere else in the country after four centuries of settlement, we have the opportunity to assure strong, statutory protection for truly vast expanses of wild land" (S. Rep. No. 96-413 1979, 374). Also at issue was the designation of wild and scenic rivers in Alaska. Because river valleys in Alaska were not extensively developed, there were extraordinary opportunities for large-scale wild and scenic river designations on federal lands in Alaska (S. Rep. No. 96-413 1979, 408). A Senate version of the Alaska lands conservation bill removed a number of the wild and scenic river designations found in a House version of the bill. In a 1979 Senate report, Senators Metzenbaum and Tsongas protested these removals and noted that "Alaska remains the last place in the nation where extensive segments of major continental rivers qualify for wild or scenic river status" (S. Rep. No. 96-413 1979, 411).

The disagreements between the Senate and the House on the Alaska lands issue threatened to extend past the five-year deadline for land withdrawals in Alaska under section 17(d)(2) of the Alaska Native Claims Settlement Act. While all of the Alaska lands withdrawn under section 17(d)(2) had also been withdrawn indefinitely under section 17(d)(1) of the Alaska Native Claims Settlement Act, the looming expiration of the section 17(d)(2) withdrawals nevertheless had symbolic political weight (Allin 2008, 237; Williss 2005, 102). The effective delaying tactics of the senators from Alaska raised the possibility that an Alaska lands conservation bill might be blocked indefinitely. Representative Udall referred to the Alaska congressional delegation's approach to negotiations over the Alaska lands conservation bill as constituting a "deliberate strategy of delay" (U.S. Congressional Record 1979, 46). This strategy of delay prevented a Senate vote on the Alaska lands conservation bill from taking place before the deadline for the expiration of the land withdrawals made under section 17(d)(2) of the Alaska Native Claims Settlement Act (Cahn 1982; Williss 2005). The Carter administration responded to this situation by taking a series of actions intended to shift the political momentum in favor of the resolution of the Alaska lands issue in Congress.

The Alaska National Interest Lands Conservation Act

President Jimmy Carter took a prominent role in promoting the cause of Alaska lands conservation on numerous occasions. In a 1977 message to Congress concerning the environment, Carter noted the extraordinary potential of the Alaska lands bill to expand the national conservation systems: "We can double the size of the Wildlife Refuge and the Park Systems, as well as add to the Forest and Wild and Scenic River Systems, at no acquisition cost" (Carter 1977, 979). Carter repeatedly urged the enactment of an Alaska lands conservation bill through a series of statements in the period 1977–80, including two environmental messages and three State of the Union messages to Congress (Carter 1977, 1978a, 1979a, 1979c, 1980a). In the 1980 State of the Union message, Carter stated that the decision on the Alaska lands conservation bill "clearly amounts to the conservation decision of the century" (Carter 1980a, 158).

President Carter was urged to take action for the protection of Alaska lands by the environmental movement and numerous members of Congress. In 1978 and 1980, the Carter administration took a series of aggressive executive actions intended to impel the passage of legislation for Alaska lands conservation in Congress (Allin 2008). In 1978, the state of Alaska filed a lawsuit in federal court in an attempt to block presidential actions to protect lands in Alaska, but this lawsuit was swiftly dismissed (Nelson 2004, 220–22; *U.S. Congressional Record* 1979, 46). In November 1978, the state of Alaska proposed state land selections that would include the transfer of millions of acres being considered for federal conservation purposes in Alaska into the ownership of the state; two days later, Interior Secretary Cecil Andrus responded by temporarily withdrawing 110.8 million acres of federal lands in Alaska for a period of three years under the 1976 Federal Land Policy and Management Act (Pub. L. 94-579) (Allin 2008; Andrus and Freemuth 2006; Miles 2009; Nelson 2004, 220–22; S. Rep. No. 96-413 1979, 133; Williss 2005). President Carter then took further actions to add a new layer of protection to some of the lands that Secretary Andrus had temporarily withdrawn in Alaska in 1978 (Allin 2008, 236). Carter took these actions under the authority of a 1906 law titled An Act for the Preservation of American Antiquities (Pub. L. 59-209, generally known as the Antiquities Act). The Antiquities Act gave unilateral authority to the president to proclaim national monuments for the protection of objects of historic or scientific interest on lands owned or

controlled by the federal government (Harmon, McManamon, and Pitcaithley 2006; Rothman 1994). In December 1978, Carter issued a series
of proclamations establishing seventeen national monuments encompassing
approximately 56 million acres in Alaska under the authority of the Antiquities Act (S. Rep. No. 96-413 1979, 133). The majority of the area of these
new national monuments in Alaska was placed under the management of the
National Park Service (H.R. Rep. No. 96-97 Part I 1979, 394). In a statement
concerning these 1978 national monument proclamations in Alaska, Carter
noted that "passing legislation to designate National Parks, Wildlife Refuges,
Wilderness Areas, and Wild and Scenic Rivers in Alaska is the highest environmental priority of my administration" (Carter 1978b, 2111). Carter characterized these new national monuments in Alaska as "the most critical areas
proposed for legislative designation—13 proposed National Parks, two proposed Wildlife Refuges, and two proposed National Forest Wilderness areas"
(Carter 1978b, 2111). Carter further stated that the new national monuments
he had proclaimed in Alaska would "remain permanent Monuments until the
Congress makes other provisions for the land" (Carter 1978b, 2111). Carter
clearly indicated his intent to encourage the enactment of the Alaska lands
bill through his national monument proclamations in Alaska and associated
actions, stating that "the actions I have taken today provide for urgently
needed permanent protections. However, they are taken in the hope that the
96th Congress will act promptly to pass Alaska lands legislation" (Carter
1978b, 2112).

The land withdrawals and national monument proclamations in Alaska by
the Carter administration together protected almost all of the areas in Alaska
under consideration for conservation by Congress (Williss 2005). While the
Alaska land withdrawals made in 1978 under the Federal Land Policy and
Management Act were temporary, the 1978 national monument designations
in Alaska were permanent unless revised or revoked by congressional or presidential action. National monument proclamations had historically served as a
politically useful form of executive protection that could set the stage for
subsequent protective legislation; indeed, a number of national parks established by Congress had initially been protected as national monuments. The
Antiquities Act allowed the president to swiftly seize the political initiative in
protecting a natural area from development during the often lengthy period
of congressional deliberation concerning statutory protections for the area
(Rothman 1994). This two-stage process of executive and legislative land
conservation would be recapitulated on a grand scale in Alaska within a
period of only two years.

The executive actions taken by the Carter administration to protect Alaska lands met with great approval from the national environmental movement, and great disapproval in Alaska. Carter's support of land conservation in Alaska was recognized by the environmental movement in 1979, when the National Wildlife Federation gave Carter the Conservationist of the Year Award (Carter 1979b). But the 1978 executive initiatives protecting lands in Alaska were also met with public protests in Alaska, unsuccessful legal actions filed by the state of Alaska and development interests in opposition to those executive initiatives, and a bill designed to revoke those executive initiatives sponsored by members of the Alaska congressional delegation (Allin 2008, 244; Andrus and Freemuth 2006; S. Rep. No. 96-413 1979, 133; The Wilderness Society 2001, 18).

The 1978 executive initiatives protecting lands in Alaska shifted the political momentum toward the passage of an Alaska lands bill by creating a new system of conservation units in Alaska that would be politically difficult for Congress or another president to dismantle. The new national monuments in Alaska protected more than 50 million acres of land from development, and these protections would be permanent unless they were revised or revoked by congressional or presidential actions. But the revocation or erosion of the national monument protections in Alaska would invite a political backlash from the national environmental movement, a movement that had demonstrated its capacity to organize a nationwide lobbying effort through the Alaska Coalition.

Threats to revoke the 1978 executive land protections in Alaska were effectively counteracted by the Carter administration. Alaska senators Stevens and Gravel sponsored a bill designed to revoke the 1978 national monument proclamations in Alaska and severely restrict the ability of the administration to apply the Antiquities Act or the Federal Land Policy and Management Act (Allin 2008, 244). Interior Secretary Andrus responded to this Stevens–Gravel bill with a threat to expand executive protections for the Alaska landscape, and the bill did not advance in Congress (Allin 2008, 245). In 1980, Andrus withdrew 40 million acres of land in Alaska to be administered as wildlife refuges and natural resource areas for a period of twenty years under the Federal Land Policy and Management Act (Allin 2008, 249; Andrus and Freemuth 2006). These new land withdrawals included 37.6 million acres of wildlife refuges that were effectively permanent under existing law, which compelled the renewal of the wildlife refuge withdrawals after the initial twenty-year withdrawal term expired. The combined effect of the 1978 national monument proclamations and the 1980 wildlife refuge withdrawals in Alaska was to give permanent executive protection to 93.6 million acres of

Alaska unless further executive or legislative actions were taken on the subject (Allin 2008, 236–49).

The final result of the prolonged political negotiations over the Alaska lands bill was a Senate bill that was designed to add approximately 105 million acres to federal conservation systems in Alaska. The Senate bill was less favorable to conservation in Alaska than the House version of the bill, but the political landscape soon shifted in favor of the Senate bill when Carter lost the 1980 presidential election to Ronald Reagan (Allin 2008). Reagan had voiced his opposition to the Alaska lands bill during his presidential campaign (Swem and Cahn 1984; Williss 2005). The Alaska lands bill therefore faced the possibility of a veto if sent to Reagan rather than Carter, and the 1980 election of a new conservative majority in the Senate meant that overcoming a veto or even further negotiations on the Alaska lands bill could prove impossible. If the Alaska lands bill was not approved, Reagan would soon be in a position to take executive actions to weaken the conservation measures of the Carter administration in Alaska (Cahn 1982, 30). Furthermore, the Alaska senators both threatened to filibuster further amendments to the Alaska lands bill (Duscha 1981; Nelson 2004). This new political circumstance meant that authorization of the House version of the Alaska lands bill was improbable (Carson and Johnson 2001; Duscha 1981). Speaking on the Alaska lands bill in 1980, Representative Millicent Fenwick of New Jersey noted that "we have to accept that which is possible even if it is not perfect" (*U.S. Congressional Record* 1980, H10528). The House of Representatives proceeded to approve the Senate version of the Alaska lands bill (Duscha 1981).

Carter hailed the enactment of the Alaska lands bill, stating that "both Houses of Congress have now endorsed the greatest land conservation legislation of the century" (Carter 1980b, 2719). Carter signed the Alaska National Interest Lands Conservation Act of 1980 (Pub. L. 96-487) into law on December 2, 1980 (Carter 1980c). The Alaska National Interest Lands Conservation Act (ANILCA) added 105.4 million acres to federal conservation systems in Alaska, including 104.1 million acres of land conservation units and 1.3 million acres of river conservation units. ANILCA further designated 56.3 million acres of wilderness across both new and preexisting conservation units in Alaska (Allin 2008, 257). The federal land management units in Alaska under ANILCA are shown in map 3.1.

ANILCA did much to serve the interests of the national environmental movement. Although ANILCA made many changes in the boundaries and management provisions of Alaska conservation units proposed by the original H.R. 39, the total acreage added to conservation systems in Alaska by ANILCA constituted approximately 92 percent of the total acreage proposed

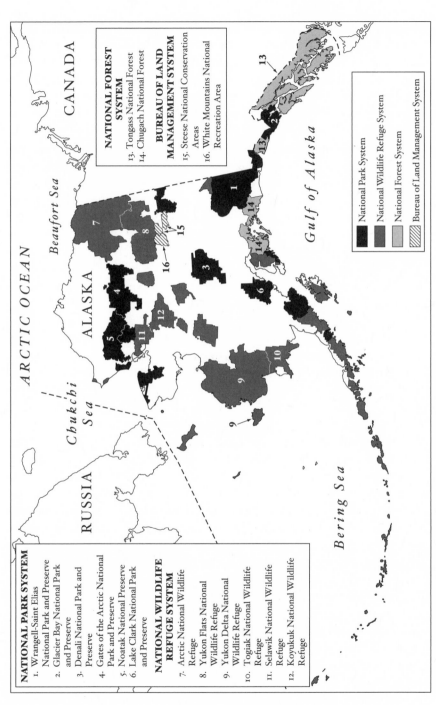

NATIONAL PARK SYSTEM
1. Wrangell–Saint Elias National Park and Preserve
2. Glacier Bay National Park and Preserve
3. Denali National Park and Preserve
4. Gates of the Arctic National Park and Preserve
5. Noatak National Preserve
6. Lake Clark National Park and Preserve

NATIONAL WILDLIFE REFUGE SYSTEM
7. Arctic National Wildlife Refuge
8. Yukon Flats National Wildlife Refuge
9. Yukon Delta National Wildlife Refuge
10. Togiak National Wildlife Refuge
11. Selawik National Wildlife Refuge
12. Koyukuk National Wildlife Refuge

NATIONAL FOREST SYSTEM
13. Tongass National Forest
14. Chugach National Forest

BUREAU OF LAND MANAGEMENT SYSTEM
15. Steese National Conservation Areas
16. White Mountains National Recreation Area

National Park System
National Wildlife Refuge System
National Forest System
Bureau of Land Management System

MAP 3.1. Federal Land Management Units in Alaska under ANILCA.

for conservation purposes by the original H.R. 39. The National Park Service and the Fish and Wildlife Service would manage approximately 95 percent of the 104.1 million acres of land conservation units established in Alaska by ANILCA, with the remaining 5 percent of the acreage of these units to be managed by the Forest Service and Bureau of Land Management (Allin 2008, 257). The vast majority of the area of the new land conservation system established in Alaska by ANILCA would therefore be managed by federal agencies that placed a high priority on nature conservation. Furthermore, large areas of Alaska would be designated as federally protected wilderness areas in which infrastructure development would generally be precluded. Indeed, ANILCA constituted the single greatest land conservation action in the history of the world up to that time (Allin 2008; Nash 2001). Carter emphasized the national conservation importance of ANILCA in remarks made during the law's signing ceremony: "We are setting aside for conservation an area of land larger than the State of California . . . we are doubling the size of our National Park and Wildlife Refuge System. By protecting 25 free-flowing Alaskan rivers in their natural state, we are almost doubling the size of our Wild and Scenic Rivers System. By classifying 56 million acres of some of the most magnificent land in our Federal estate as wilderness, we are tripling the size of our Wilderness System" (Carter 1980c, 2756).

The 1980 legislative protections for lands in Alaska under ANILCA superseded the various executive protections for lands in Alaska previously established by the Carter administration in the period 1978–80 (94 Stat. 2487). ANILCA marked the end of the critical period for Alaska lands conservation that had begun ten years before. The policy image established during this critical period represented a complex compromise between conservation, development, recreation, hunting, and subsistence interests. This policy image supported a vast expansion of federal conservation units across Alaska, the continuation of customary and traditional subsistence activities in some areas of federal conservation units in Alaska, the continuation of mechanized access to federal conservation units in Alaska for traditional uses, the continuation of sport hunting in some areas of federal conservation units in Alaska, and a number of accommodations for natural resource extraction in Alaska (including support for the continuation of preexisting mining claims in conservation units, logging in the Tongass National Forest, and oil development in northern Alaska). This policy image therefore contained numerous concessions to interests other than environmental interests, and indeed ANILCA protected a considerably smaller area in Alaska than environmental interests had sought. Some of the most significant concessions to development interests in ANILCA were found in the law's wilderness designations. The 56 million acres of wilderness designated in Alaska by ANILCA constituted approximately 38 percent of the wilderness acreage proposed in the original H.R. 39 (Allin 2008, 228, 257). Furthermore, ANILCA authorized a number of uses for wilderness in Alaska that were generally not permitted in wilderness areas outside of Alaska, including subsistence activities and mechanized access for traditional uses (Norris 2002; The Wilderness Society 2001). Despite these compromises made for the purposes of subsistence and mechanized access, the wilderness areas established in Alaska would serve the fundamental conservationist aim of precluding major infrastructure or industrial development in those areas. The policy image and institutional arrangements that emerged from this critical period therefore established a large-scale conservation system in Alaska that was protected with concessions to allow for the extensive subsistence activities and mechanized transportation that were essential elements of the traditional way of life of communities in many parts of Alaska.

The authorization of ANILCA reflected a pattern of predominantly supportive political and media attention to nature conservation in Alaska during the Alaska lands conservation punctuation. These patterns of attention are examined further in the following section.

Congressional and Media Attention

This study finds evidence of a prevailing rise in political and media attention in favor of expanding nature conservation areas in Alaska during the Alaska lands conservation punctuation. Summaries of all congressional hearings and *New York Times* articles addressing the topic of wilderness in Alaska were collected and coded for this study, covering the period 1959–84 (beginning with Alaska statehood and encompassing the period during which the idea of large-scale land conservation in Alaska was debated and eventually approved). The keyword wilderness was chosen for data collection purposes because that keyword was generally used in congressional and media discussions of nature conservation in Alaska. To assess patterns of support for nature conservation in Alaska over time, the topics in these congressional hearings and news articles were coded positive or negative in tone toward the expansion of nature conservation areas in Alaska. Topics coded positive were supportive of expanding nature conservation areas in Alaska, while topics coded negative were not supportive of expanding nature conservation areas in Alaska.

The Alaska lands conservation punctuation was accompanied by a sharp increase in congressional and media attention in support of expanding nature conservation areas in Alaska. Figure 3.1 shows the annual number of congressional hearing topics coded positive or negative in tone toward expanding nature conservation areas in Alaska in the period 1959–84. Because some hearings included multiple distinct topics, the number of topics coded is larger than the number of hearings.

The early years of this record (1959–69) show little attention to the topic of wilderness in Alaska in congressional hearings. The Alaska lands conservation punctuation was marked by a modest rise in positive tone topics

FIGURE 3.1. Topics in U.S. congressional hearings on Alaska and wilderness (1959–84) coded positive or negative in tone toward the expansion of nature conservation areas in Alaska. Topics coded positive were supportive of expanding nature conservation areas in Alaska; topics coded negative were not supportive of expanding nature conservation areas in Alaska.

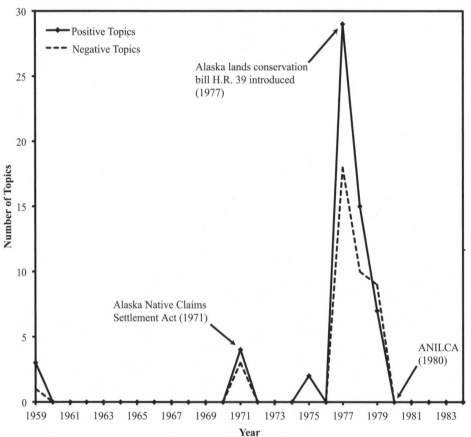

supportive of expanding nature conservation areas in Alaska discussed in congressional hearings in 1971 (the year that the conservation planning provisions of the Alaska Native Claims Settlement Act were approved); a lull from 1972 to 1976, during which time Interior Secretary Morton withdrew lands in Alaska for conservation planning purposes and Alaska lands legislation stalled in Congress; and a sharp rise in positive tone topics in 1977 (the year that the ambitious H.R. 39 bill for Alaska lands conservation was introduced in Congress). Positive tone topics remained a prominent but declining feature of congressional hearings on this issue until 1980, when attention to this issue in congressional hearings ceased. However, the data also show substantial evidence of sustained congressional opposition to Alaska nature conservation during the punctuation. The punctuation featured a modest rise in negative tone topics that were not supportive of expanding nature conservation areas in Alaska discussed in congressional hearings in 1971, a sharp rise in negative tone topics in 1977, and a declining but still prominent trend in negative tone topics until 1980. While the number of positive tone topics outweighed the number of negative tone topics on this issue from 1971 through 1978, negative tone topics prevailed in 1979. The data therefore indicate a predominantly positive wave of congressional attention supporting the expansion of nature conservation areas in Alaska during the punctuation examined in this chapter, but also indicate a smaller wave of negative congressional attention as a major feature of this punctuation (a dynamic that reflects the prolonged controversy over Alaska lands conservation legislation in Congress).

Positive attention to the expansion of nature conservation areas in Alaska during the Alaska lands conservation punctuation is evident not only in data from congressional hearings, but also in data from newspaper articles in the *New York Times*. Figure 3.2 shows the annual number of *New York Times* article topics coded positive or negative in tone toward expanding nature conservation areas in Alaska in the period 1959–84. One major topic concerning nature conservation in Alaska was coded for each *New York Times* article.

The early years of this media record (1959–77) show a pattern of limited, sporadic, but usually positive attention supportive of the expansion of nature conservation areas in Alaska. A sharp rise in positive tone articles was evident in 1978, the year that H.R. 39 first passed in the House of Representatives and the year of major executive actions protecting lands in Alaska by the Carter administration. A peak in negative tone articles occurred in 1979 as the controversy over Alaska lands conservation legislation persisted, followed by a peak of positive tone articles with the authorization of ANILCA in 1980. The data indicate a predominantly positive wave of media attention to the

Figure 3.2. Topics in *New York Times* articles on Alaska and wilderness (1959–84) coded positive or negative in tone toward the expansion of nature conservation areas in Alaska. Article topics coded positive were supportive of expanding nature conservation areas in Alaska; article topics coded negative were not supportive of expanding nature conservation areas in Alaska.

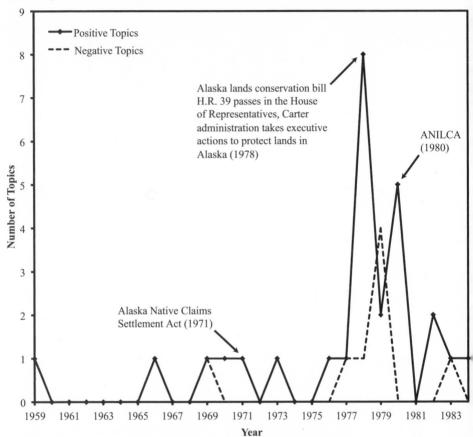

topic of expanding nature conservation areas in Alaska during the punctuation examined in this chapter, but also indicate a smaller wave of negative media attention as a major feature of this punctuation. Overall, the number of positive tone article topics outweighed the number of negative tone article topics in the media record during the Alaska lands conservation punctuation. These patterns of media attention are similar to the patterns of congressional attention examined earlier in this chapter, although attention to this issue in the *New York Times* was limited and tended to lag behind congressional attention.

Overall, the record of congressional hearings and *New York Times* articles indicates a sharp rise in political and media attention that was predominantly supportive of the expansion of nature conservation areas in Alaska during the Alaska lands conservation punctuation. These findings are consistent with the punctuated equilibrium theory (Baumgartner and Jones 2009). However, both records also show evidence of a significant wave of negative attention opposing the expansion of nature conservation areas in Alaska during the punctuation. These data reflect the fact that the Alaska lands conservation punctuation was a prolonged and highly controversial period of policy reform.

The Institutional Legacies of ANILCA for Environmental Protection in Alaska

The equilibrium period following the authorization of ANILCA was characterized by the progressive development of federal programs for conservation, recreation, and subsistence management across the conservation units established by ANILCA (Miles 2009). The magnificent scenery and wildlife values protected by ANILCA provided extraordinary opportunities for outdoor recreation and established an expanded basis for the development of a thriving tourism industry in Alaska (Carson and Johnson 2001). By precluding infrastructure development across large regions in Alaska, the ANILCA equilibrium allowed vast areas of the Alaska landscape to remain largely undisturbed by human activities. The largely undisturbed character of the ANILCA conservation units served the ecological preservation goals of the environmental movement and also provided a valuable natural laboratory for scientific research on those lands (Shah 2005). However, ANILCA also accommodated natural resource development activities that would pose continuing challenges to conservation in some areas of Alaska; the example of logging in the Tongass National Forest of Alaska is considered next.

The Tongass National Forest

A particularly consequential compromise in favor of Alaska development was contained in provisions of ANILCA supporting logging in the Tongass National Forest of Alaska, the largest national forest in the United States and part of the largest contiguous temperate rain forest ecosystem in the world (Durbin 2005, 4; H.R. Rep. No. 95-1045 Part I 1978, 158). Forest Service management of the Tongass constituted a massive federal subsidization of a local timber economy (H.R. Rep. No. 95-1045 Part I 1978, 159). First, the Forest Service pursued road building projects in the Tongass that aided private timber operations and also improved surface transportation for the communities in the Tongass (Durbin 2005; Ross 2000). Second, the Forest Service sold Tongass timber to industry at rates that constituted egregious examples of Forest Service subsidization of private timber operations on federal lands (H.R. Rep. No. 96-97 Part I 1979, 501; Durbin 2005). In a 1979 Senate report, Senators Metzenbaum and Tsongas noted that the cost to the federal government of building roads in the Tongass and administering Tongass timber sales was much higher than the return to the federal government from these timber sales: "The difference is paid by the Federal taxpayer to provide Alaskan timber jobs. Timber programs are run at a loss in other National Forests as well, but nowhere is the loss as high as the Tongass" (S. Rep. No. 96-413 1979, 395).

Environmental concerns in the Tongass received little attention from either the Forest Service or the timber industry, and decades of timber cutting in the Tongass caused significant damage to its temperate rain forest ecosystem. The pulp mills in the Tongass emitted large amounts of air and water pollution, systematically violating federal air and water pollution control laws (Durbin 2005). Conservation efforts in the Tongass were pursued by the environmental group Southeast Alaska Conservation Council, which was formed by local community members in the Tongass region. The Southeast Alaska Conservation Council integrated its efforts to protect the Tongass forest into the larger Alaska Coalition effort to establish a comprehensive lands conservation bill for Alaska. But while ANILCA established more than 5 million acres of wilderness areas in the Tongass that would be closed to further development, it also served to reinforce the existing arrangements under which the Forest Service subsidized timber operations in other areas of the Tongass (Allin 2008, 260). The Tongass provisions in ANILCA supported continued logging in the Tongass with a minimum of $40 million in annual federal funding and a legislated target for a high volume of timber cutting in the forest. By reserving some areas in the Tongass with limited forest resources as wilderness but leaving other Tongass areas with extensive forest resources open to subsidized timber operations, the Tongass provisions of ANILCA enabled more logging in the Tongass both on Forest Service lands and Alaska Native lands (Durbin 2005, 100–105; Nelson 2004; Ross 2000).

Following the approval of ANILCA, the Southeast Alaska Conservation Council joined forces with other environmental organizations to lobby for additional protections for the Tongass. The environmentalist campaign for the Tongass led to incremental progress for Tongass conservation through the approval of the 1990 Tongass Timber Reform Act. The Tongass Timber Reform Act repealed ANILCA's annual appropriation and timber-cutting target for the Tongass, and designated more than 1 million additional acres to be protected from logging in the Tongass. While the Tongass Timber Reform Act did not serve to end logging in the Tongass, logging in the Tongass diminished over time as the two pulp mills in the Tongass that were the most important sources of demand for Tongass timber closed due to market forces and mounting concerns over the environmental implications of their logging and pulp operations. The ANILCA equilibrium in the Tongass was therefore characterized by a gradual progression from a regional economy emphasizing natural resource extraction to a regional economy emphasizing tourism that depended on natural resource conservation (Durbin 2005, 200–202, 312–15; Nelson 2004, 258–60; Ross 2000, 260–64). As discussed in the next section, the ANILCA equilibrium made conservation a central purpose of national policy in Alaska.

Conservation in Alaska following ANILCA

The ANILCA equilibrium posed significant management challenges for federal agencies with jurisdiction over the new conservation units in Alaska. In the aftermath of ANILCA, federal agencies such as the National Park Service built up their Alaska regional organizations to support their new land management responsibilities in Alaska (Miles 2009). The continued operation of preexisting mining sites within the conservation units established by ANILCA posed challenges for visitation and conservation in some of the units, as did the reclamation of abandoned mining sites (Ross 2000; Shah 2005). Substantial agency effort was also required to regulate subsistence activities, sport hunting, and mechanized access in ANILCA conservation units. Furthermore, ANILCA mandated the use of a number of advisory councils designed to provide sustained input from Alaska citizens concerning the management of subsistence activities on federal lands in their state (Norris 2002; The Wilderness Society 2001).

ANILCA gave the National Park Service jurisdiction over a number of areas in Alaska designated as national preserves in which sport hunting was allowed. These national preserves reflected the interest of sport hunters in preserving hunting opportunities in the new conservation units in Alaska (H.R. Rep. No. 95-1045 Part I 1978, 359–60). While the National Park Service had historically attempted to exclude hunting from the national parks, in 1974 Congress established two new national preserves (Big Thicket National

Preserve in Texas and Big Cypress National Preserve in Florida) in which sport hunting would be allowed under the jurisdiction of the National Park Service (S. Rep. No. 95-1300 1978, 334). These first national preserves created a new administrative category for the National Park Service that would be applied on a large scale in Alaska by ANILCA. The sport hunting provisions of ANILCA required a greater adjustment by the National Park Service than by the Fish and Wildlife Service, as the Fish and Wildlife Service had generally accepted hunting in the National Wildlife Refuges elsewhere in the nation (Swem and Cahn 1984).

ANILCA also allowed mechanized access for traditional uses in conservation units in Alaska (Williams 1997). Mechanized access to conservation areas in Alaska was an allowance of great importance for subsistence activities and sport hunting due to the immense size and limited road network of the state (H.R. Rep. No. 95-1045 Part I 1978, 268, 359). A statement by three members of Congress in a 1978 House of Representatives report noted that "the airplane is a way of life in Alaska, given the vast distances" (H.R. Rep. No. 95-1045 Part I 1978, 360). Yet ANILCA only authorized mechanized access to Alaska conservation units for a limited range of uses, such as traditional activities in those units (The Wilderness Society 2001, 43). For example, the National Park Service excluded the use of snowmobiles from the former area of Mount McKinley National Park (a park unit that ANILCA had enlarged and renamed Denali National Park and Preserve) because the use of snowmobiles was not a traditional activity in Mount McKinley National Park prior to ANILCA (The Wilderness Society 2001, 46).

In sum, ANILCA created significant management challenges for conservation in Alaska. ANILCA nevertheless constituted a conservation action of great national and international importance, as discussed next.

National and International Impacts of the Alaska Lands Conservation Punctuation

The Alaska lands conservation punctuation was of singular importance in the history of U.S. nature conservation. ANILCA established Alaska as the lead state in American land conservation by giving Alaska the majority of the total land areas in the National Park System, National Wildlife Refuge System, and National Wilderness Preservation System (Andrus and Freemuth 2006; Williams 1997). Furthermore, the system of land conservation units established by ANILCA in Alaska generally differed from the system of land conservation units found elsewhere in the nation both in scale and interconnection. In the contiguous United States, the capacity of protected natural areas to conserve wildlife populations was often constrained by the limited acreages and scattered distribution of these protected areas, with the habitat and ranges of wildlife populations frequently extending far outside the boundaries of these conservation units. The limitations of the land conservation system in the contiguous United States reflected a diverse and fragmented series of political efforts at conservation that had occurred in competition with widespread development pressures (Fischman 2003; Runte 2010). Extensive development between scattered conservation units had caused widespread habitat losses and created barriers to wildlife migration in the contiguous United States. By contrast, ANILCA was the result of a comprehensive conservation planning effort. The ANILCA conservation units were generally very large and often interconnected (Hunter and Wood 1981). The ANILCA conservation units were notable for protecting intact ecosystem processes on large scales, including the wildlife migrations that were a major feature of the natural world in Alaska (Duffy et al. 1999; Swem and Cahn 1984). The extended process of coordinated conservation planning that occurred during the Alaska lands conservation punctuation thus led to a system of protected natural areas in Alaska that typically conserved larger intact areas of habitat and posed fewer barriers to wildlife movements than the system of land conservation units in the contiguous United States (Marchetti and Moyle 2010). This conservation accomplishment of ANILCA reflected the emphasis on ecosystem and wildlife preservation that was an important element of the Alaska lands conservation debate (Swem and Cahn 1984).

The unprecedented scale of the ANILCA conservation actions constituted a landmark event for the international conservation movement. The Alaska lands conservation punctuation was of particular international importance because some of the areas conserved by ANILCA contained habitats used by migratory wildlife, including birds migrating to and from many parts of the world (Cahn 1982). The conservation importance of ANILCA was reinforced by the creation of conservation units in Canada that connected across the border to conservation units in Alaska, creating two large transnational conservation areas along the border between Alaska and Canada. Both of these large transnational protected areas represented the culmination of international conservation proposals that predated ANILCA by decades (Kaye 2006, 38–39; Williss 2005, 8–9). These transnational conservation areas were assembled in stages over time. In 1942, Canada created a Kluane national park reserve on the border with Alaska, a conservation action that had been suggested by the U.S. government (Theberge 1978). Subsequent reforms beginning in 1972 strengthened protections for the Kluane park and created a combined conservation area consisting of Kluane National Park and Reserve and Kluane Wildlife Sanctuary in Canada; these contiguous conservation units are referred to collectively as the Kluane conservation area in this book (Parks Canada 2010; Theberge 1978). In 1980, ANILCA established a complex of both new and enlarged conservation units (including the Wrangell-Saint Elias National Park and Preserve, Glacier Bay National Park and Preserve, and Tetlin National Wildlife Refuge) that interconnected not only with each other but also with the Kluane conservation area (Hunter and Wood 1981; Matz 1999). This exceptionally large transnational conservation area was further enlarged by the 1993 creation of Tatshenshini-Alsek Park by the Canadian province of British Columbia (Matz 1999; Tatshenshini-Alsek Management Board 2001). In northern Alaska, ANILCA established an enlarged Arctic National Wildlife Refuge and created the Yukon Flats National Wildlife Refuge; these two interconnected wildlife refuges were subsequently joined across the border by new national parks in the Yukon territory of Canada. Canada established Ivvavik National Park in 1984 and the adjoining Vuntut National Park in 1995, which interconnected with each other and with the Arctic National Wildlife Refuge and Yukon Flats National Wildlife Refuge in the United States; these four contiguous conservation units together formed another large transnational protected area (Parks Canada 2007; Kaye 2006, 211; Rennicke 1995; Woods 2003).

These two transnational protected areas gave Canada and the United States a pronounced interest in conservation matters across their shared border. Of particular concern to the two nations was the Porcupine caribou herd that migrated seasonally between the Arctic National Wildlife Refuge in the United States and adjoining areas in Canada (Rennicke 1995; Woods 2003). The overall effect of U.S. and Canadian land conservation efforts in and around Alaska was the creation of a vast system of northern conservation units that protected wildlife migrations and other ecosystem processes by establishing large protected areas with extensive interconnections (Marchetti and Moyle 2010). The U.S. federal and Canadian national parks and wildlife conservation areas found in and around Alaska are shown in map 3.2.

An important design feature of the conservation areas interconnecting across the border between Alaska and Canada was the authorization of subsistence activities by indigenous peoples in those areas. ANILCA authorized subsistence activities on the U.S. side of these transnational conservation areas, and similar reforms authorized subsistence activities on the Canadian side of these areas (Kopas 2007; Parks Canada 2007, 2010; Rennicke 1995; Stevens 1997; Theberge 1978). In both Alaska and northern Canada, the

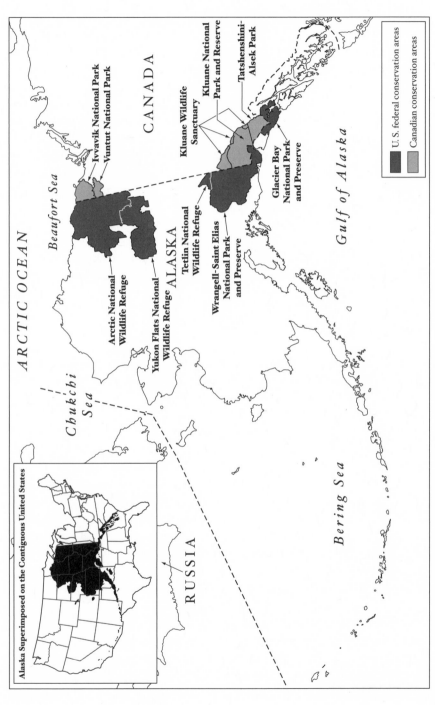

MAP 3.2. U.S. Federal and Canadian National Parks and Wildlife Conservation Areas in and around Alaska.

establishment of large conservation units therefore often allowed for the continuation of traditional subsistence activities in those units by native communities (and institutional arrangements were established to allow for local input into subsistence management decisions). The integration of native subsistence activities into large conservation areas in Alaska and northern Canada represented a significant break with the long history of conservation areas that prohibited inhabitation or subsistence activities by native peoples who had previously relied on natural resources in those protected areas (Parks Canada 2010; Spence 1999; Stevens 1997). This new approach to conservation in North America reflected an acknowledgment of the value of traditional indigenous cultures and lifestyles, and an acknowledgment that the ecosystems placed in conservation units had in many cases supported subsistence by native peoples for thousands of years (Fagan 2004; Stevens 1997). The integration of indigenous subsistence activities into northern conservation areas in Alaska and Canada thereby preserved interacting cultural and ecological processes that had co-existed for millennia.

Summary

The Alaska lands conservation punctuation began with the 1970 addition of Alaska conservation planning provisions into an early version of legislation designed to settle Alaska Native land claims. Systematic conservation planning for Alaska was subsequently approved as part of the Alaska Native Claims Settlement Act, and that law thereby constituted a defining institutional arrangement for both the Alaska pipeline punctuation and the Alaska lands conservation punctuation. The Alaska lands conservation punctuation continued through prolonged congressional debate and aggressive executive initiatives on this subject, and ended with the 1980 authorization of a large-scale system of conservation units in Alaska under ANILCA. The policy image guiding ANILCA represented a complex effort to support an unprecedented expansion of protected natural areas in Alaska while also continuing support for subsistence activities and sport hunting across some of these protected areas, mining in preexisting claims in these protected areas, mechanized access for traditional uses in these protected areas, and natural resource development in Alaska. ANILCA formed the decisive institutional arrangement of the Alaska lands conservation punctuation, and that institutional arrangement has led to enduring policy consequences. For more than three decades, the conservation system established by ANILCA has endured as the largest area of protected lands found in any state in America.

The institutional arrangements and policy image established in the Alaska lands conservation punctuation contained a complicated compromise that distributed benefits among environmental interests, subsistence interests, sport hunting interests, and various Alaska development interests. The tremendous size and extensive interconnections of ANILCA conservation units and adjoining conservation units in northern Canada allowed wildlife migrations and ecosystem processes to continue largely undisturbed across vast areas, thereby serving the conservation interests of the environmental movement. In turn, the abundant wildlife populations protected by these conservation units provided the basis of sustained subsistence activities and sport hunting that could be pursued through mechanized access in some protected areas under the terms of ANILCA. Further accommodations were made in ANILCA for the benefit of mining, timber, and oil development in Alaska. In an approach to policy design with similarities to the Alaska pipeline punctuation, the Alaska lands conservation punctuation established institutional arrangements that broadly distributed the benefits of Alaska's natural resources. Unlike the Alaska pipeline punctuation, the Alaska lands conservation punctuation constituted a sweeping victory for the U.S. environmental movement.

The critical period examined in this chapter established the protection of wilderness and wildlife as central purposes of national policy in Alaska. But the national intent to protect the wilderness qualities of Alaska would subsequently be confronted by a massive oil spill in the coastal waters of Alaska. As shown in the next chapter, this environmental disaster led to a critical period of reform with enduring consequences for the marine oil trades of Alaska, the United States, and the world.

The *Exxon Valdez* Disaster and the Oil Pollution Act of 1990

IN 1977 THE TRANS-ALASKA PIPELINE SYSTEM commenced oil shipments, thereby establishing a marine oil trade that posed significant hazards to the coastal environment of Alaska. The risk to the environment
posed by the marine oil trade in Alaska had been widely discussed during the
Alaska pipeline punctuation. In the course of the Alaska pipeline punctuation,
the Nixon administration and the oil industry had made repeated assurances
concerning the safety of the oil tanker fleet to be used in the Trans-Alaska
Pipeline System. Despite these assurances, oil tankers in Alaska operated with
few safeguards against marine oil pollution for more than a decade after the
Trans-Alaska Pipeline System began oil shipments. This pattern of complacency came to an abrupt end in 1989, when the oil tanker *Exxon Valdez*
spilled approximately 11 million gallons of North Slope oil into the waters of
coastal Alaska (Alaska Oil Spill Commission 1990). The *Exxon Valdez* disaster triggered a critical period of reform that led to a major realignment in
marine oil pollution policy at the national level. The *Exxon Valdez* oil spill
punctuation had sweeping consequences for the management of the marine
oil trade of Alaska.

The equilibrium period between the authorization of the Trans-Alaska
Pipeline System in 1973 and the *Exxon Valdez* disaster in 1989 was characterized by incremental policy changes in both federal law and Alaska state law

concerning marine oil pollution. During this equilibrium period, the federal role in marine oil pollution policy was authorized by a fragmented collection of laws. These laws included the 1972 Federal Water Pollution Control Act (Clean Water Act), the 1972 Ports and Waterways Safety Act, the 1973 Trans-Alaska Pipeline Authorization Act, the 1974 Deepwater Port Act, and the 1978 Outer Continental Shelf Lands Act Amendments. The federal role in this policy domain fell under the scope of the commerce clause of the U.S. Constitution, as oil tankers crossed both interstate and international borders. State laws also played a considerable role in marine oil pollution policy during this equilibrium period. Although some in the oil industry and Congress claimed that federal law should preempt state law in this policy domain under the supremacy clause of the U.S. Constitution, Alaska and a number of other states had nevertheless enacted marine oil pollution laws (Ramseur 2010).

The question of the extent to which state laws should be allowed to coexist with federal laws in this policy domain was tested during this equilibrium period through a series of lawsuits. The general boundaries of federal preemption in marine oil pollution law were tested in two major cases that reached the U.S. Supreme Court, while the specific boundaries of federal preemption over Alaska state law in this policy domain were tested in a federal district court and federal circuit court of appeals. In the first Supreme Court case (*Askew v. American Waterways Operators, Inc.*), oil shippers claimed that federal law preempted a Florida state law imposing unlimited liability for state and private damages resulting from oil spills in Florida waters. In 1973, the Supreme Court found that the Florida law in question was not preempted by federal law, noting that the relevant federal law itself contained language disclaiming federal preemption of state law concerning marine oil pollution (Alaska Oil Spill Commission 1990). However, a subsequent Supreme Court decision in the case *Ray v. Atlantic Richfield Co.* limited the reach of state laws addressing the operation of oil tankers in state waters. In this case, oil industry and oil shipping interests claimed that federal law preempted a Washington state law establishing oil tanker design and operation standards that were more stringent overall than federal standards. This Washington state law had been enacted in 1975 in part due to concerns that shipments from the Trans-Alaska Pipeline System to oil refineries in the Puget Sound region would threaten the coastal environment of Washington. In 1978, the Supreme Court found that elements of the Washington state law in question were preempted by the Ports and Waterways Safety Act of 1972, a federal law that the court interpreted as establishing uniform standards for the design of oil tankers nationwide (Alaska Oil Spill Commission 1990). These two Supreme Court cases therefore upheld the principle of a joint system of federal and state management of marine oil pollution, but also held that state efforts to regulate oil tanker design and construction were preempted by federal law.

In 1976, the state of Alaska enacted a marine oil pollution law including requirements for oil tanker design and tug escort vessels (tugboats that would accompany tankers in transit). In the federal district court case *Chevron v. Hammond*, this Alaska state law was challenged by the oil industry on the grounds that federal law preempted the state law. Following the 1978 Supreme Court decision in *Ray v. Atlantic Richfield Co.*, the litigants in *Chevron v. Hammond* agreed that the tanker design and tug escort requirements of the Alaska law were preempted by federal law. Additional preemption rulings in this case were followed by the repeal of some parts of Alaska state law concerning marine oil pollution (Alaska Oil Spill Commission 1990). The partial repeals of Alaska state law in this policy domain and the preemption agreements and findings in *Chevron v. Hammond* diminished but did not extinguish the role of the state of Alaska in regulating the marine oil trade of Alaska. The state of Alaska appealed a ruling on the issue of state ballast water discharge regulations, and in 1984 a federal circuit court of appeals found that Alaska state ballast water discharge regulations were not preempted by federal law (Alaska Oil Spill Commission 1990). Alaska state law also maintained an unlimited liability standard for marine oil spills (Millard 1993).

The fragmented collection of federal and state laws dealing with marine oil pollution in the United States was the subject of sustained criticism and attempts at reform by environmentalists and some members of Congress during the equilibrium period preceding the *Exxon Valdez* disaster. These efforts at marine oil pollution reform were largely defeated during this equilibrium period due to disagreements in Congress and opposition from the oil and shipping industries (Alcock 1992; Kurtz 2004; Millard 1993). A notable oil spill occurred in the marine oil trade of Alaska during this equilibrium period. In 1987, the tanker *Glacier Bay* ran aground in Cook Inlet and spilled an estimated 150,000 gallons of North Slope oil into the waters of Alaska; this relatively small oil spill constituted approximately 1 percent of the volume of oil that would be later spilled by the tanker *Exxon Valdez* into the waters of Alaska in 1989 (Carr 1992). The relatively small oil spill from the *Glacier Bay* in Alaska failed to create the political momentum needed to achieve major reforms in the marine oil trades of Alaska or the United States.

Due to the general lack of progress in marine oil pollution policy during this equilibrium period, the marine oil transportation system in Alaska employed only limited safeguards against the hazard of oil spills at sea. The design standards for oil tankers required only the minimal protection of a single hull separating the oil cargo tanks from the sea; more protective hull designs that reduced the risk of oil spills were available but not required by law (Alcock 1992). The *Exxon Valdez* was a single-hull tanker (Alaska Oil Spill Commission 1990, 123). The tracking of oil tankers along the coast of Alaska was constrained by limited radar coverage, and course errors by oil tankers might therefore go undetected in many areas of Alaska. Oil tankers in Alaska also sailed largely without the protection of tug escort vessels, which could be used to correct course errors by the tankers, fight fires on the tankers, and respond to oil spills. Oil tankers in Alaska were vulnerable to severe weather conditions, but gaps in the regional weather reporting network limited the availability of weather warnings. The marine oil transportation system of Alaska did not have a comprehensive environmental monitoring and research program either to guide oil spill response efforts, or to assess the environmental impacts of oil spills on the coast of Alaska. The oil spill response equipment available in the marine oil transportation system of Alaska was insufficient to meet the demands of a major oil spill at sea (Alaska Oil Spill Commission 1990).

The capabilities of some of the safeguards against marine oil pollution in Alaska declined prior to the *Exxon Valdez* disaster. The power of the radar used for vessel tracking in Prince William Sound was reduced, the oil spill response team in the Sound was disbanded, and oil spill response equipment was not maintained in a state of readiness (Alaska Oil Spill Commission 1990; Busenberg 2008). The Alyeska contingency plan for a marine oil spill in Prince William Sound was essentially designed to manage a small oil spill, but did not contain a realistic plan for managing a large oil spill (Alyeska 1987; Clarke 1993, 1999). The Alyeska contingency plan reflected the view of the oil industry that a large marine oil spill in Prince William Sound was a highly unlikely event—a complacent perspective that failed to account for the massive environmental risks inherent in the use of oil tankers to transport North Slope oil. Clarke (1999) has characterized the 1987 Alyeska oil spill contingency plan for Prince William Sound as a fantasy document that served as a symbolic reassurance of system safety rather than as an effective practical approach for managing a large oil spill. The fundamental inadequacies of this contingency plan were starkly revealed in 1989 when the *Exxon Valdez* oil spill swiftly overwhelmed both the oil spill response resources and the organizational structures that were available to respond to the spill. In sum, the marine oil trade of Alaska possessed oil pollution safeguards with limited capabilities (and in some cases declining capabilities) in the years preceding the *Exxon Valdez* disaster (Busenberg 2008; Clarke 1993; EVOSTC 2004; PWS RCAC 1993d, 2009c; PWS Science Center 2004). This pattern of complacency was broken by a critical period of policy reform triggered by the 1989 *Exxon Valdez* oil spill.

The *Exxon Valdez* Disaster

The chain of events leading to the *Exxon Valdez* oil spill began when the tanker *Exxon Valdez* departed from the Valdez oil terminal with a cargo of North Slope oil from the Trans-Alaska pipeline one evening in 1989. The captain of the *Exxon Valdez* ordered a course change to avoid ice in the waters of Prince William Sound. The site of the oil terminal in Prince William Sound had been chosen to reduce the risks posed by marine ice to oil tankers, but significant marine ice hazards were still present in the region due to a tidewater glacier that frequently discharged ice into the waters of Prince William Sound (The Wilderness Society, Environmental Defense Fund, and Friends of the Earth 1972). Tankers in Prince William Sound normally traveled along designated routes (referred to as traffic lanes) that were designed to avoid navigational hazards, but ice in the water could drift into those traffic lanes and pose a hazard to the tankers. This marine ice hazard was particularly acute at night when the ice was difficult or impossible to see. The captain of the *Exxon Valdez* ordered a departure from the traffic lanes to avoid the ice. The captain then left the bridge under the control of other members of the crew after giving instructions to change course to reenter the traffic lanes once the *Exxon Valdez* was clear of the ice. The crew members in control of the tanker subsequently failed to change course in time to avoid grounding on Bligh Reef in Prince William Sound. The grounding on Bligh Reef punctured most of the cargo tanks of the *Exxon Valdez*. Subsequent investigations found that the *Exxon Valdez* disaster was caused by crew error, and that alcohol impairment and crew fatigue contributed to the disaster (Alaska Oil Spill Commission 1990; NTSB 1990).

The *Exxon Valdez* spill revealed the inadequacies of the systems for oil spill prevention and response in Prince William Sound. The navigational error that led to the *Exxon Valdez* disaster was not detected by the U.S. Coast Guard radar tracking system in Prince William Sound. There was no tug escort vessel in position to aid the *Exxon Valdez* in a course correction at the time of the navigational error or to assist in oil spill response after the tanker grounded. The oil spill response equipment then available in Prince William Sound was not ready for swift deployment to contain the spill, and that equipment did not have the capacity to recover and store the volume of oil spilled. Alternative response methods were tested on the *Exxon Valdez* oil spill (including the use of dispersants designed to break up the spilled oil in the water and attempts to burn the spilled oil), but these tests accomplished little. A storm subsequently dispersed the oil spilled from the *Exxon Valdez*. Approximately 11 million gallons of oil from the *Exxon Valdez* spilled into the sea and spread across more than one thousand miles of Alaskan coastline, causing massive wildlife casualties. The 1989 *Exxon Valdez* disaster was the largest marine oil spill in the history of America up to that time (Alaska Oil Spill Commission 1990; Burger 1997; Busenberg 2008; Clarke 1993; EVOSTC 2004, 2009; NTSB 1990; PWS RCAC 1993d, 2009c).

The *Exxon Valdez* disaster acted as a focusing event, focusing public and political attention on the environmental hazards of the marine oil trade (Birkland 1997; Birkland and Lawrence 2001). As described next, this heightened attention marked the beginning of a critical period of reform in marine oil pollution policy.

The Oil Pollution Act of 1990

Prior to the *Exxon Valdez* disaster, congressional efforts to enact comprehensive oil spill reform legislation had long been stalled. The U.S. attorney general completed a study of the oil spill issue in 1975, which led to a bill for oil spill reform introduced in Congress in 1976 (Millard 1993). The House of Representatives and the Senate passed different versions of comprehensive oil spill legislation but did not reach an agreement to resolve their differences concerning the legislation prior to 1989. Congress had therefore failed to enact comprehensive oil spill legislation prior to 1989 despite more than a decade of reform efforts in this policy domain (Millard 1993; Randle 2012). The *Exxon Valdez* disaster ended this equilibrium period and triggered a critical period of reform that brought strong public and political support for the longstanding efforts of environmental interests to strengthen regulation of the marine oil trade (Alcock 1992; Grumbles and Manley 1995; Kurtz 2004; Millard 1993; Ramseur 2010). Reflecting on the long history of attempts to pass oil spill reform in Congress, Representative Don Young of Alaska stated that: "We all know that the *Exxon Valdez* catastrophe was the engine that finally drove this bill to completion and I am dismayed that Alaska had to suffer that experience before this Congress decided to take decisive action" (*U.S. Congressional Record* 1990, H6944). Reflecting on the public impact of the *Exxon Valdez* spill, Representative Bruce Vento of Minnesota stated that "the catastrophic size of the spill and its effects on the pristine qualities of Alaska's environment captured national attention and outraged the American public" (*U.S. Congressional Record* 1989, H7974).

Following the *Exxon Valdez* disaster, oil spill legislation was given further momentum in Congress by three smaller oil spills that occurred along the U.S. coastline in June 1989. Although these additional marine oil spills in

1989 were not disasters on the scale of the *Exxon Valdez* spill, they provided timely evidence that marine oil spills were not uncommon events along the American coastline. Representative Walter Jones of North Carolina noted the effect of the *Exxon Valdez* disaster and other oil spills of 1989 in giving new momentum to comprehensive oil spill reform efforts in Congress after years of unsuccessful prior attempts to achieve such a reform: "More than once during these years of frustration, it was said that it would take a major oil spill disaster to achieve enactment of this bill. Unfortunately, this has proved true. The catastrophe of the *Exxon Valdez* and the tragedies of the other oilspills this summer have made this bill an exceedingly popular cause" (*U.S. Congressional Record* 1989, H7955).

The need for comprehensive federal reform on the subject of marine oil pollution policy was made clear by this string of oil spill disasters, as noted in 1989 by Representative Norman Lent of New York:

> During 1989 this Nation has seen dramatic evidence of the need for comprehensive oilspill legislation, beginning with the terrible oilspill in Alaska on March 24, 1989. That spill was followed by the three spills in Rhode Island, the Houston Ship Channel, and the Delaware River, on the weekend of June 24, 1989. Up until now we have been operating under a patchwork of Federal, State, and local laws and regulations to clean up oilspills and to pay for damages. Clearly the existing prevention measures and removal capabilities of the combined authorities are inadequate to clean up spills and to respond to the liability and compensation problems (*U.S. Congressional Record* 1989, H7962).

A 1989 Senate report found that at least five different federal statutes dealt with oil spill liability and compensation (S. Rep. No. 101-94 1989). Reform efforts during the *Exxon Valdez* oil spill punctuation focused on creating an overarching federal law that would strengthen and coordinate regulations for marine oil spill planning, prevention, response, liability, and compensation in the United States. Major issues in the congressional debate over oil spill reform during the critical period included questions of jurisdiction over marine oil spills, liability standards for marine oil spills, new design standards for oil tankers, and special provisions to enhance the environmental safety of the marine oil trade in Alaska.

Prior to 1989, a central point of disagreement in the prolonged congressional debate concerning comprehensive oil spill reform was the issue of whether such a federal reform should preempt state laws in this policy domain (Randle 2012). The preemption controversy had proven to be a major obstacle to comprehensive oil spill reform in Congress. By 1989, a number of states had enacted laws and funds designed to deal with marine oil pollution along their coastlines. A 1989 Senate report found that twenty-four states had enacted oil spill compensation and liability laws, and also found that seventeen of these states had enacted oil spill laws without specified limits to the liability of those responsible for the oil spills (S. Rep. No. 101-94 1989). These state laws set a precedent that supported a non-preemptive federal approach in this policy domain, as did the 1973 Supreme Court ruling in the case *Askew v. American Waterways Operators, Inc.*, which had rejected the federal preemption of Florida state oil pollution law (Alaska Oil Spill Commission 1990).

Environmental groups supported reform that would perpetuate the authority of states to impose their own oil spill liability and response standards on the basis that joint federal and state regulation of marine oil spills would provide stronger protections against oil spills than federal regulation alone (Millard 1993; Ramseur 2010). Oil and shipping interests supported federal preemption in this policy domain on the basis that joint federal and state regulation of marine oil spills created a complex and risky regulatory environment for oil shipping. The oil tanker trade involved tanker transits in the maritime jurisdictions of many states; in the absence of federal preemption in this policy domain, oil tanker spills could fall under numerous and differing state response and liability regimes. Furthermore, the unlimited liability for oil spills found in many state laws translated into unlimited financial risk for any corporation that spilled oil within the reach of those laws (Millard 1993; S. Rep. No. 101-94 1989). Oil and shipping interests argued that if the states were allowed to continue enforcing their own separate oil spill liability standards, the resulting liability exposure for oil spills could disrupt or even shut down the marine oil trade of America (Grumbles and Manley 1995). But the fact that the marine oil trade had continued in the maritime jurisdictions of states that had established unlimited liability standards for marine oil pollution indicated that reforms perpetuating state liability standards for marine oil pollution would not disrupt or shut down the marine oil trade of America (S. Rep. No. 101-94 1989). Indeed, the policy image that prevailed in the congressional debate focused on the potential for stronger liability standards at both the state and federal levels to encourage safety on the part of corporations involved in marine oil shipping by increasing the financial risks of oil spills for those corporations. During the 1989 Senate debate over oil spill reform, Senator Max Baucus of Montana stated: "Spills are too much an accepted cost of doing business for the oil shipping industry. Many in the industry seem to have decided that it is cheaper to spill and pay for its cleanup than it is to prevent spills and develop effective techniques to contain them.

Consequently, our legislation encourages prevention by imposing new, tougher penalties for those who pollute or fail to comply with contingency plans. S. 686 also ensures that States can retain and enact unlimited liability laws to encourage responsible industry behavior" (*U.S. Congressional Record* 1989, S9691).

Senator Frank Lautenberg of New Jersey noted that the *Exxon Valdez* spill "proved that our existing oilspill response system is wholly inadequate. The liability limits are too low, making it cheaper to spill than to take the necessary precautions. Contingency plans often are not worth the paper they are written on" (*U.S. Congressional Record* 1989, S9702). Senator John Chafee of Rhode Island also noted that the oil spill reforms under consideration in the Senate provided "very strong incentives for companies to take every precaution possible to prevent a spill" (*U.S. Congressional Record* 1989, S9692). The *Exxon Valdez* disaster prompted the end of the longstanding congressional stalemate over preemption in oil spill reform (Millard 1993). The prevailing congressional view on the issue of state laws and funds concerning marine oil pollution was summarized in a joint statement by ten members of Congress in a 1989 House of Representatives report: "The public has an interest in preserving these state funds and laws, and that public interest is far more compelling to us than the oil and shipping industry's narrow interest in preempting them" (H.R. Rep. No. 101-242 Part 2 1989, 150). The arguments for federal preemption in oil spill liability and response law were overridden in the congressional debate (Randle 2012; *U.S. Congressional Record* 1989, H7971–72). A prominent element of the argument against federal preemption in this policy domain was the role played by the state of Alaska oil spill liability law in the *Exxon Valdez* oil spill response; in 1989, Representative George Miller of California noted that "Alaska's tough liability law is a major reason why Exxon spent nearly $2 billion in this year's cleanup effort" (*U.S. Congressional Record* 1989, H7972).

The most far-reaching reform established during the *Exxon Valdez* oil spill punctuation was a mandate for a double-hull design standard for oil tankers operating in U.S. maritime jurisdiction. In a double-hull oil tanker design, the bottom and side sections of the inner hull were enclosed within a second outer hull designed to absorb impacts. An internal space that did not contain oil separated the protective outer hull from the inner hull, while the oil cargo was enclosed within the inner hull. The double-hull tanker design therefore provided a substantial additional layer of protection for the oil tanks when compared to the conventional single-hull oil tankers that constituted the majority of the tanker fleets in the marine oil trades of both Alaska and the world at the time of the *Exxon Valdez* oil spill punctuation (Alaska Oil Spill

Commission 1990, 123; Alcock 1992, 104). With a protective outer hull on the bottom and sides of the tanker, the double-hull tanker design provided protection against both vessel collisions (which were expected to occur on the side of the vessel) and groundings (which were expected to occur on the bottom of the vessel). This double-hull tanker design was one of several tanker designs (including the double-bottom design and double-sided design) that had been proposed for the purpose of reducing marine oil pollution. A double-bottom tanker design had an outer hull on the bottom but not the sides of the tanker, and was therefore expected to provide protection against groundings but not against collisions with other vessels. A double-sided tanker design had an outer hull on the sides but not the bottom of the tanker, and was therefore expected to provide protection against collisions with other vessels but not against groundings (Alcock 1992; Eyres 2007; *U.S. Congressional Record* 1989, S9710). The double-hull, double-bottom, and double-sided tanker designs are collectively referred to as protective-hull designs in this book.

Protective-hull designs for tankers had been debated at both the national and international levels for many years prior to the *Exxon Valdez* disaster. The 1972 Ports and Waterways Safety Act authorized the U.S. Department of Transportation to regulate tanker construction and operation (Alcock 1992). This regulatory authority was given to the U.S. Coast Guard within the Department of Transportation. The Coast Guard proposed regulations for protective-hull design standards for tankers, but subsequently withdrew those proposed regulations in the face of sustained opposition from oil and shipping interests. Oil and shipping interests opposed standards for protective-hull designs based on the argument that these protective-hull designs would be more expensive to build than the traditional single-hull designs. This sustained industry opposition succeeded in greatly delaying national and international standards for protective-hull designs on oil tankers (Alcock 1992).

The international nature of the marine oil trade suggested a need for an international approach to the issue of oil tanker design standards. The United States unsuccessfully proposed a worldwide double-bottom design standard for oil tankers at both the 1973 and 1978 international conferences that produced the International Convention for the Prevention of Pollution from Ships and a protocol to that convention (Alcock 1992). While these conferences did not produce an agreement for a global double-bottom standard for oil tankers, they did produce a compromise agreement under which segregated ballast tanks would be protectively located on some large new oil tankers. Under this agreement, segregated ballast tanks designed to carry only water would be built between some of the oil tanks and the outer surfaces of the ships. Since segregated ballast tanks never carried oil, an accident that breached a ballast tank would not spill oil. But segregated ballast tanks could not protect most of the oil tanks in a tanker, due to the limited number of ballast tanks needed for safe navigation (Tan 2006). Therefore, a single-hull tanker design with protectively located ballast tanks provided less protection for the cargo tanks of the vessel when compared to a double-hull tanker design. The vulnerability of this partial protection was vividly demonstrated by the 1989 *Exxon Valdez* oil spill, as the single-hull tanker *Exxon Valdez* had protectively located ballast tanks that did not serve to prevent this spill (Alaska Oil Spill Commission 1990, 123; Alcock 1992; NTSB 1990, 85–86; Randle 2012, 33).

During the Alaska pipeline punctuation, the Nixon administration had offered assurances that double-bottom tankers would be used in the Alaska oil trade (thereby reinforcing the argument for the safety of the marine component of the Trans-Alaska Pipeline System when compared to a Trans-Canada pipeline that required no tankers). These assurances proved to be symbolic and did not subsequently translate into a regulatory mandate for protective-hull oil tanker designs in the marine oil trade of Alaska. In 1989 the majority of the tankers allowed to operate in Alaska were single-hull vessels (Alaska Oil Spill Commission 1990, 123; Alcock 1992, 104).

The equilibrium period preceding the *Exxon Valdez* disaster was therefore marked by only incremental progress on the issue of protective-hull designs for oil tankers. The combined pressure of Coast Guard proposals to require protective hulls on tankers, efforts by the United States to promote international requirements for double-bottom oil tankers, and lawsuits by environmental organizations meant to promote the adoption of protective-hull standards for oil tankers in America had been met with sustained opposition from the oil and shipping industries (Alcock 1992). The overall result was incremental policy progress in the form of an international agreement for the introduction of protectively located ballast tanks on some oil tankers—a tanker design that offered significantly less protection against oil spills when compared to a double-hull tanker design. However, this equilibrium period also produced a series of studies examining the question of whether protective-hull vessel designs did in fact offer greater protection against cargo spills when compared to single-hull designs. These studies provided considerable evidence indicating that protective-hull designs were an effective approach to reducing oil pollution at sea, and these studies would play an important role in the *Exxon Valdez* oil spill punctuation (*U.S. Congressional Record* 1989, S9710). Substantial operational experience with protective-hull designs accumulated over time due to national and international regulations mandating double hulls on vessels carrying chemicals or liquefied flammable gas, and also due to voluntary efforts by industry to build some oil tankers with protective-hull designs (*U.S. Congressional Record* 1989, S9705–6). Therefore, the debate over protective-hull designs for oil tankers during the *Exxon Valdez* oil spill punctuation was informed by considerable prior study, international negotiations, and operational experience (Alcock 1992).

Environmental interests that had long supported protective-hull designs for oil tankers used the *Exxon Valdez* oil spill punctuation to press that cause. For example, the environmental group Natural Resources Defense Council called for protective hulls on oil tankers in congressional hearings during this punctuation (Millard 1993). By contrast, oil and shipping interests opposed a protective-hull requirement for oil tankers and cast doubt on the efficacy of protective-hull designs in preventing marine oil pollution (Alcock 1992; Millard 1993). Arguments in opposition to a protective-hull mandate consisted of various claims concerning the increased cost, fire risk, explosion risk, and vessel salvage difficulties associated with protective hulls. These arguments were all undermined by operational experience with protective-hull designs on vessels carrying oil, liquefied natural gas, and hazardous chemicals. This operational experience indicated that tankers with protective hulls were neither prohibitively expensive for industry nor unsafe due to fire, explosion, or salvage issues. This operational experience furthermore indicated the efficacy of protective hulls in preventing cargo spills from tankers (*U.S. Congressional Record* 1989, S9704–8). The oil industry nevertheless called for further study of protective-hull designs during the *Exxon Valdez* oil spill punctuation, and some members of Congress called for further regulatory rule-making concerning protective-hull designs for oil tankers. However, these calls for further study and regulatory rule-making were undermined not only by operational experience with protective-hull designs but also by the availability of completed studies indicating the effectiveness of protective hulls in preventing cargo spills and by the failure of past regulatory rule-making efforts to establish protective-hull requirements (Millard 1993; *U.S. Congressional Record* 1989, S9704–13). Senator Brock Adams of Washington supported a congressional mandate for double hulls on new oil tankers, noting that the issue of protective hulls on oil tankers had been "studied to death" (*U.S. Congressional Record* 1989, S9705). The arguments for a double-hull oil tanker mandate were further reinforced by a 1990 oil spill from the tanker *American Trader* off the coast of California, as it was argued that the *American Trader* spill could have been prevented by a double hull (Alcock 1992; Millard 1993).

The arguments against protective hulls and calls for further study of protective hulls were overridden in the congressional debates, and the national reforms established during the *Exxon Valdez* oil spill punctuation included a mandate for the progressive introduction of double-hull oil tankers into the maritime jurisdiction of America.

The *Exxon Valdez* oil spill punctuation produced a number of reforms focused specifically on issues of marine oil pollution in Alaska, with the Alaska congressional delegation playing a prominent role in these reform efforts. In the critical periods of policy reform previously examined in chapters 2 and 3 of this book, elected officials from Alaska often acted in opposition to the environmental movement. By contrast, the *Exxon Valdez* oil spill punctuation was marked by a general alignment of the interests of elected officials from Alaska and the national environmental movement. The *Exxon Valdez* oil spill had caused widespread and highly visible damage to wildlife, ecosystems, and scenery across large areas of coastal Alaska. The *Exxon Valdez* oil spill had also caused severe economic and social disruptions in local communities in the spill region (Alaska Oil Spill Commission 1990; Bushell and Jones 2009). The *Exxon Valdez* disaster revealed that the traditional support for natural resource development by elected officials from Alaska did not extend to the point of tolerating major environmental disasters induced by industry in Alaska. Following the *Exxon Valdez* oil spill, the Alaska state government enacted and implemented reforms designed to strengthen the environmental management of the marine oil trade along the coast of Alaska. The Alaska congressional delegation unanimously supported oil spill reform at the national level and worked to ensure that this reform included provisions specific to Alaska that would strengthen the environmental management of the marine oil trade in coastal Alaska. These provisions reflected the particular attention paid by the Alaska congressional delegation to the environmental concerns and initiatives of Alaskans.

Community members from the region surrounding the Valdez oil terminal had long voiced concerns about the environmental risks of the marine oil trade of Alaska (The Wilderness Society, Environmental Defense Fund, and Friends of the Earth 1972). In 1987, a group from a community in the Prince William Sound region asked Alyeska to consider forming a citizen advisory group designed to provide advice to Alyeska on environmental concerns; Alyeska initially declined but reconsidered in 1989 following the *Exxon Valdez* disaster (Ginsburg, Sterling, and Gotteherer 1993). A Regional Citizens' Advisory Council formed in Prince William Sound in 1989. In 1990, this new advisory council signed a contract with Alyeska that guaranteed the council independence from Alyeska, access to Alyeska facilities, annual funding from

Alyeska, and a continuation of the contract as long as the Alaska pipeline continued transporting oil (Ginsburg, Sterling, and Gotteherer 1993; PWS RCAC 1991a).

The Regional Citizens' Advisory Council in Prince William Sound was modeled on advisory groups in the Shetland Islands of Scotland that provided advice to the government and industry organizations managing the Sullom Voe oil and gas terminal in the Shetland Islands (Ginsburg, Sterling, and Gotteherer 1993). The Sullom Voe terminal was managed in a collaborative partnership between industry and the local community, providing an international model that received substantial attention in Alaska following the *Exxon Valdez* disaster (Bohlen 1993). However, the approach used to manage the Sullom Voe terminal was not directly transferable to Alaska due to fundamental differences in the governance structures found in Alaska and the Shetland Islands. In 1974, the United Kingdom had enacted legislation granting the local government of the Shetland Islands (the Shetland Islands Council) sweeping authority over development and environmental conservation on the coastline of the Shetland Islands. The Shetland Islands Council negotiated an equal partnership with the oil industry for the development and management of the Sullom Voe terminal. The Shetland Islands Council therefore functioned as a local government that not only co-managed the Sullom Voe terminal in an equal partnership with the oil industry, but that also had regulatory authority over the Sullom Voe terminal and the harbor activities of tankers in the Shetland Islands. The Shetland Islands Council thereby had a far greater degree of influence over the management of the Sullom Voe terminal than did local governments and interest groups in Prince William Sound. There was no equal partnership between Alyeska and local governments for the management of the Valdez oil terminal, and regulatory authority over the marine oil terminal and oil tankers in Prince William Sound was held by state and federal agencies rather than by local governments. Nonetheless, the Shetland experience provided an instructive model of local community involvement in the environmental management of a major oil terminal and oil tanker operations (Ginsburg, Sterling, and Gotteherer 1993).

The idea of using citizen advisory councils to oversee the environmental management of the marine oil trade of Alaska was incorporated into U.S. law through the efforts of the Alaska congressional delegation during the *Exxon Valdez* oil spill punctuation. During the 1989 Senate debate on oil spill legislation, Senator Frank Murkowski of Alaska proposed an amendment designed to allow local citizen input into the environmental management of the marine oil trade of Alaska in both Prince William Sound and Cook Inlet. Murkowski noted the role of government and industry complacency in the *Exxon Valdez* disaster and proposed a mandate for two advisory councils in Alaska, inspired by the Shetland experience, to counteract this problem:

> In an effort to combat this complacency, my amendment involves local citizens, the people who have the most to lose in the event of an oil spill, in the decision making and monitoring process in an advisory capacity. While regulators and industry representatives may grow complacent over time, the people whose way of life depends on the environmentally safe operation of the terminal will not. . . . The idea behind this legislation is not original. It is patterned after a successful system of citizen involvement in the Shetland Islands—at the Sullom Voe terminal. Because of the effective partnership that has developed at Sullom Voe, Sullom Voe is considered the safest crude oil terminal in Europe . . . One of the underlying causes of the *Exxon Valdez* was that the people charged with the responsibility of overseeing the safe transportation of oil in Prince William Sound had become complacent. The best way to combat this complacency is to involve local citizens in the process in an advisory capacity. This system works in the Shetland Islands and it can work here. Only when local citizens are involved in the process will we begin to build the trust necessary to change the present system from confrontation to consensus (*U.S. Congressional Record* 1989, S9904).

The national oil spill reforms enacted by Congress in 1990 included provisions mandating two advisory councils designed to allow local citizen oversight of the environmental management of the marine oil trade of Alaska. These councils would subsequently play prominent roles in contributing to the progressive improvement of environmental safeguards in the marine oil trade of Alaska (Busenberg 2008).

The *Exxon Valdez* disaster propelled congressional action to conclude the longstanding debate over oil spill reform with a new Oil Pollution Act. The *U.S. Congressional Record* and congressional committee reports concerning the debate over the Oil Pollution Act demonstrated widespread concern among members of Congress over the environmental disaster caused by the *Exxon Valdez* oil spill and sweeping support for reform of federal oil pollution law in response. This pattern of strong congressional support for oil pollution reform in the aftermath of the *Exxon Valdez* disaster was clearly evident in the voting for the final bill; the Oil Pollution Act of 1990 was enacted by unanimous votes in both the House of Representatives and the Senate (Randle 2012). President George Bush had also voiced concern over the problem of marine oil pollution in the aftermath of the *Exxon Valdez* disaster. Shortly after the *Exxon Valdez* disaster, the Bush administration issued a White House fact sheet stating that "the oil spill in Prince William Sound, simply put, is one of the greatest environmental tragedies in American history" (Bush 1989, 386). This fact sheet also stated that the president urged Congress to consider and act upon oil spill legislation promptly (Bush 1989). Bush supported an international approach to marine oil pollution regulation, but this international approach was largely rejected by Congress in favor of a unilateral approach (Grumbles and Manley 1995). Faced with overwhelming congressional support for oil spill reform, President Bush signed the Oil Pollution Act of 1990 (Pub. L. 101-380) into law on August 18, 1990 (Bush 1990).

The approval of the Oil Pollution Act of 1990 (OPA 90) ended the brief critical period of reform that had commenced with the *Exxon Valdez* oil spill in 1989. OPA 90 established a major realignment of the marine oil trade of America that provided a new consolidated framework of federal law in this policy domain (Ramseur 2010). OPA 90 strengthened the federal liability standards for marine oil spills, and allowed each state to impose its own liability and response requirements for marine oil pollution within its jurisdiction. OPA 90 required the progressive introduction of double-hull oil tankers into the maritime jurisdiction of the United States, a process that would occur in stages up to 2015. By 2015, OPA 90 required double hulls on all oil tankers operating within the maritime jurisdiction of the United States; this jurisdiction included the exclusive economic zone claimed by America, a zone that generally extended 200 nautical miles seaward from the U.S. coastline but that was truncated in some areas by international maritime boundaries (Reagan 1983a, 1983b; Sohn et al. 2010). OPA 90 also strengthened oil spill contingency planning requirements for all vessels and facilities involved in the marine oil trade of America, established new requirements designed to reduce intoxication and fatigue among oil tanker crews (including requirements for alcohol screening and restrictions on work hours), and created a new consolidated federal response fund to support federal efforts to respond to marine oil spills (Beaver, Butler, and Myster 1994; National Research Council 1998; Randle 2012; Wilkinson, Pittman, and Dye 1992).

OPA 90 also included provisions specific to Alaska. OPA 90 required a major reinforcement of oil spill response resources in Prince William Sound and the deployment of a vessel tracking system with increased range in Prince William Sound. OPA 90 also established a requirement for three organizations designed to provide advice and information concerning the environmental management of the marine oil trade in Alaska, including two citizens' advisory councils and a research organization. Specifically, OPA 90 established a requirement for a Regional Citizens' Advisory Council (RCAC) for the

Prince William Sound and *Exxon Valdez* oil spill regions, a requirement ful-
filled by the preexisting Prince William Sound RCAC. OPA 90 also established a requirement for a RCAC for the nearby Cook Inlet region. The
Prince William Sound Regional Citizens' Advisory Council (PWS RCAC)
would be directed by representatives of local communities and interest groups
in the region affected by oil spilled in the *Exxon Valdez* disaster. The Cook
Inlet Regional Citizens Advisory Council (CIRCAC) would be directed by
representatives of local communities and interest groups affected by the oil
trade in the Cook Inlet region. Map 4.1 shows the Prince William Sound and
Cook Inlet regions.

OPA 90 mandated that both RCACs would receive annual funding from
the oil industry in their respective regions. The oil industry would not be
represented in the membership of the RCACs, and OPA 90 required that the
RCACs operate as self-governing organizations (CIRCAC 2004; PWS RCAC
1991a, 2009c). OPA 90 also established a research organization named the
Prince William Sound Oil Spill Recovery Institute (OSRI), which would
receive funding from a federal trust and would support research and development projects on environmental issues and oil spill response (OSRI 2003).

In addition to the organizations required by OPA 90, a number of other
new organizations with environmental protection missions emerged in Alaska
following the *Exxon Valdez* disaster. In 1989, the Prince William Sound
(PWS) Science Center was established in Alaska as a research organization
that focused on environmental issues and oil spill response in Prince William
Sound (Bohlen 1993; PWS Science Center 2004). In 1989, Alyeska established a Ship Escort/Response Vessel System to serve as a dedicated oil spill
prevention and response organization for Prince William Sound (Alyeska
2009). In 1989, the federal government and the state of Alaska formed an
intergovernmental trustee council to oversee damage assessment and natural
resource restoration efforts associated with the *Exxon Valdez* disaster (GAO
1993). In 1991, this trustee council was transformed into the *Exxon Valdez*
Oil Spill Trustee Council (referred to here as the Trustee Council) under the
terms of an agreement by which Exxon settled state and federal claims arising
from the *Exxon Valdez* disaster (GAO 1993). The Trustee Council would
subsequently operate as a joint federal-state intergovernmental organization
overseeing the use of civil settlement funds resulting from the *Exxon Valdez*
disaster (EVOSTC 2004, 2009; GAO 1993; Hunt 2009). In 1990, a collection of state and federal agencies established an intergovernmental organization named the Joint Pipeline Office to coordinate their efforts to manage the
Trans-Alaska pipeline (McBeath et al. 2008). The critical period of reform
triggered by the *Exxon Valdez* disaster therefore established a collection of

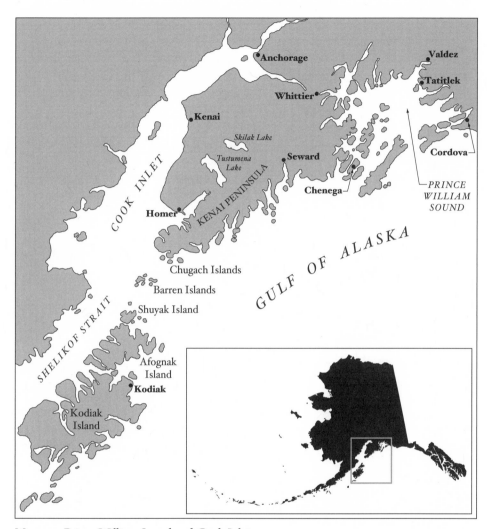

MAP 4.1. Prince William Sound and Cook Inlet.

new organizations focused on various aspects of oil spill prevention, response, and restoration in Alaska.

Overall, the *Exxon Valdez* oil spill punctuation established a wide range of new institutional arrangements designed to improve the safeguards against oil pollution in the United States. The policy image supporting these institutional arrangements reflected a concerted effort to reduce the risk of marine oil pollution nationwide through the progressive introduction of oil tankers with protective-hull designs into the maritime jurisdiction of the United States, to encourage safe practices in marine oil transportation by authorizing a marine oil pollution liability system that increased the financial risk posed by marine oil spills to industry, to enhance oil spill preparedness nationwide, to reinforce the system of marine oil spill prevention and response in Alaska, and to establish a set of new organizations designed to maintain and improve the safeguards against oil pollution in Alaska over time.

OPA 90 contained some concessions to oil and shipping interests. OPA 90 established stricter federal liability standards for oil spills, but unlimited federal liability under OPA 90 was reserved for certain cases such as oil spills due to negligence. OPA 90 also established a lengthy process for the gradual replacement of single-hull oil tankers with double-hull oil tankers in the maritime jurisdiction of the United States, so the existing industry fleet of single-hull oil tankers would not have to be replaced swiftly. However, in most respects the *Exxon Valdez* oil spill punctuation was a victory for environmental interests that had long sought stronger environmental regulation of the marine oil trade. The *Exxon Valdez* oil spill punctuation also reflected an intense wave of media and political attention in support of marine oil pollution reform, as shown in the next section.

Congressional and Media Attention

This study finds evidence of a pronounced rise in political and media attention supportive of stronger regulation of marine oil pollution during the *Exxon Valdez* oil spill punctuation. Summaries of all congressional hearings and *New York Times* articles including the topic of oil spills or the topic of oil pollution were collected and coded for this study, covering the period 1981–99. These data were not specific to Alaska but rather national in scope, since the *Exxon Valdez* oil spill punctuation was not focused solely on Alaska but instead marked a period of national reform with consequences for oil transportation throughout the maritime jurisdiction of the United States (although this period of reform had particularly sweeping consequences for the marine oil trade of Alaska). To assess patterns of support for oil pollution reform, the topics in these congressional hearings and news articles were coded positive or negative in tone toward stronger regulation of marine oil pollution. Topics coded positive were supportive of stronger regulation of marine oil pollution, while topics coded negative were not supportive of stronger regulation of marine oil pollution.

The *Exxon Valdez* oil spill punctuation was accompanied by a sharp increase in congressional and media attention in support of stronger regulation of marine oil pollution. Figure 4.1 shows the annual number of congressional hearing topics coded positive or negative in tone toward stronger

Figure 4.1.

Topics in congressional hearings on oil spills or oil pollution (1981–99) coded positive or negative in tone toward stronger regulation of marine oil pollution. Topics coded positive were supportive of stronger regulation of marine oil pollution; topics coded negative were not supportive of stronger regulation of marine oil pollution.

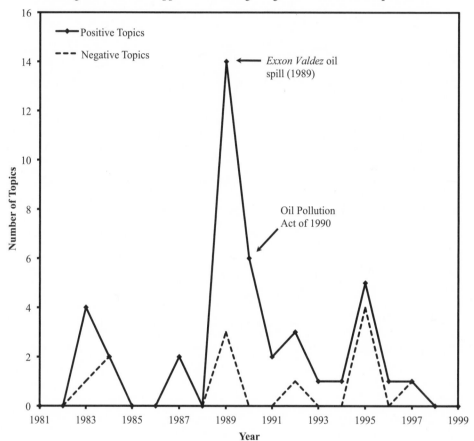

regulation of marine oil pollution in the period 1981–99. Because some hearings included multiple distinct topics, the number of topics coded is larger than the number of hearings.

A sharp rise in positive tone topics supportive of stronger regulation of marine oil pollution is evident in congressional hearings in 1989, the year of the *Exxon Valdez* disaster. The number of positive tone topics outweighed the number of negative tone topics for most of the years in the data set.

Data from newspaper articles in the *New York Times* during the *Exxon Valdez* oil spill punctuation show a highly pronounced wave of positive attention supportive of stronger regulation of marine oil pollution. Figure 4.2 shows the annual number of *New York Times* article topics coded positive or negative in tone toward stronger regulation of marine oil pollution in the period 1981–99. One major topic concerning regulation of marine oil pollution was coded for each *New York Times* article.

The *New York Times* record shows a prevailing pattern of positive attention supportive of stronger regulation of marine oil pollution, with an intense wave

Figure 4.2. Topics in *New York Times* articles on oil spills or oil pollution (1981–99) coded positive or negative in tone toward stronger regulation of marine oil pollution. Article topics coded positive were supportive of stronger regulation of marine oil pollution; article topics coded negative were not supportive of stronger regulation of marine oil pollution.

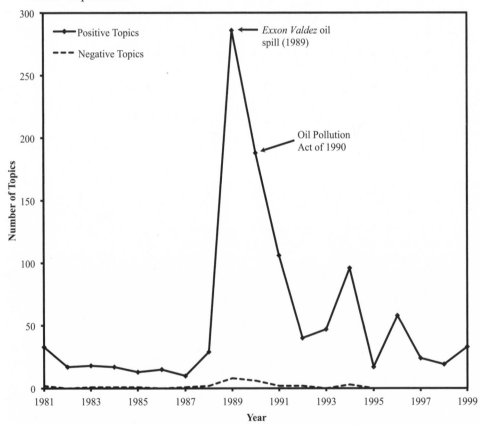

of positive attention evident in 1989 and 1990. Positive attention in this media record peaked in 1989, reflecting the extensive media coverage of the *Exxon Valdez* disaster (Birkland and Lawrence 2001). The number of positive tone article topics greatly outweighed the number of negative tone article topics in the *New York Times* record during the *Exxon Valdez* oil spill punctuation.

Overall, the congressional hearings and *New York Times* articles indicate a sharp rise in political and media attention that was predominantly supportive of stronger regulation of marine oil pollution during the punctuation examined in this chapter. These findings are consistent with the punctuated equilibrium theory (Baumgartner and Jones 2009). The enduring consequences of the *Exxon Valdez* oil spill punctuation for environmental protection in Alaska are examined next, followed by an examination of the national and international impacts of this punctuation.

The Enduring Impacts of the *Exxon Valdez* Oil Spill Punctuation in Alaska

The new institutional arrangements established during the *Exxon Valdez* oil spill punctuation subsequently contributed to a wide range of incremental advances in environmental protection in the marine oil trade of Alaska and the general region of the *Exxon Valdez* spill. One major advance in the marine oil trade of Alaska has been the implementation of the national double-hull oil tanker standard. Double-hull oil tankers have progressively replaced single-hull oil tankers in the marine oil trade of Alaska under the requirements of OPA 90 (Alcock 1992; CIRCAC 2000, 2001; National Research Council 1998; PWS RCAC 1995c, 2002c, 2003c, 2004b, 2005b, 2011a). Some of the new double-hull oil tankers in the marine oil trade of Alaska are designed with additional redundant safeguards against marine oil spills. These redundantly designed tankers include two hulls, two propellers, two rudders, and two engine rooms on each vessel. The redundant propellers, rudders, and engine rooms are designed to decrease the risk of a loss of vessel steering or propulsion that might lead to a marine oil spill (PWS RCAC 2002c, 2003c, 2004b). By reducing the risk of oil spills at sea, these redundantly designed tankers also provide oil and shipping corporations with protection against the elevated financial risks facing those corporations due to oil spill liability reforms following the *Exxon Valdez* disaster (Beaver, Butler, and Myster 1994; Busenberg 2008; Kurtz 2004). The value of the double-hull design standard for the safety of the marine oil trade in Alaska was demonstrated in 2006, when a double-hull oil tanker laden with oil grounded in Cook Inlet; while the outer hull of the tanker was cracked in the grounding, the inner hull of the tanker remained intact, and no oil was spilled from the tanker (PWS RCAC 2006, 2007b).

The capabilities of vessel tracking systems have been greatly expanded in Prince William Sound and Cook Inlet since 1989. These vessel tracking systems can be used to detect navigational problems so that tankers can correct those problems before they lead to oil spills through tanker collisions or groundings. OPA 90 specifically mandated the deployment of a vessel tracking system with increased range in Prince William Sound. In response to this mandate, in 1995 the U.S. Coast Guard deployed a new vessel tracking system in Prince William Sound with an expanded tracking range that covered all of Prince William Sound (by comparison, the range of the prior vessel tracking system did not cover most of Prince William Sound). This expanded tracking range was accomplished by equipping oil tankers with satellite-based global positioning system (GPS) transponders, which relayed the position of the tankers to the Coast Guard. The result was a highly accurate system for the continual tracking of oil tanker movements throughout Prince William Sound (PWS RCAC 1995a, 1995c). The new vessel tracking system in Prince William Sound set an early precedent for the widespread adoption of satellite-based vessel tracking in Alaska. In 2004, the U.S. Coast Guard mandated that satellite-based vessel tracking transponders be installed on oil tankers in Cook Inlet, allowing the coast guard to track oil tanker vessel movements throughout Cook Inlet as well (Busenberg 2008; PWS RCAC 2007a).

The weather reporting systems in Prince William Sound and Cook Inlet have also been significantly reinforced since 1989. Severe weather conditions pose significant hazards to oil tanker navigation in the coastal regions of Alaska, and weather reporting constitutes an important element of navigational safety in Alaska. In 1993, the PWS RCAC conducted a survey of oil tanker officers that indicated a need for better weather reporting in Prince William Sound. In the same year the PWS RCAC proposed the deployment of new weather reporting equipment to fill gaps in the weather monitoring system for Prince William Sound, with the goal of reducing the risks posed by severe weather to oil tankers in that region (Busenberg 2008; PWS RCAC 1993a, 1994a, 1994b, 1995a, 1995c; PWS Science Center 2004). The PWS RCAC proposal was supported by the oil industry, the Alaska Department of Environmental Conservation (ADEC), and the U.S. Coast Guard. In response, two members of Alaska's congressional delegation sought federal funding for the equipment. With the support of federal funding administered through the U.S. National Oceanic and Atmospheric Administration (NOAA), new weather reporting equipment was deployed by the U.S. Coast Guard at four sites in Prince William Sound in 1995 (PWS RCAC 1993b, 1994b, 1994c, 1995a). Additional weather reporting stations were subsequently installed in Prince William Sound through various collaborative projects involving the PWS RCAC, the U.S. Coast Guard, OSRI, the PWS Science Center, and local communities (OSRI 2003; PWS RCAC 2003b, 2004a, 2009b, 2009c). The weather reporting system in Cook Inlet has also been reinforced since 1989, due to a collaborative research and development project involving NOAA, the U.S. Coast Guard, and the CIRCAC that examined the safety and efficiency of navigation in Cook Inlet and Prince William Sound. In 1998 and 1999, the project led to the deployment of new weather reporting equipment at three sites in Cook Inlet (Busenberg 2008).

Significant progress in marine ice detection has occurred in coastal Alaska since the *Exxon Valdez* disaster. Marine ice poses hazards to oil tankers in both Prince William Sound and Cook Inlet. Indeed, marine ice has damaged tankers in both regions (Alaska Oil Spill Commission 1990; CIRCAC 1999; NTSB 1990; PWS RCAC 1995c). Before 1989, marine ice monitoring in the Inlet and Sound occurred periodically through vessel reports and satellite imagery. In 1993, the PWS RCAC conducted a survey of oil tanker officers that indicated a need for better marine ice reporting in Prince William Sound (PWS RCAC 1993a, 1994a). In particular, the Columbia Glacier (a tidewater glacier in Prince William Sound) was generating numerous icebergs that drifted into the tanker lanes of Prince William Sound. The PWS RCAC, OSRI, and PWS Science Center therefore collaborated to support iceberg monitoring research on the Columbia Glacier. The PWS RCAC also studied a variety of systems for improved marine ice detection in collaboration with NOAA, the Canadian Coast Guard, the oil industry, and an engineering firm (NOAA 1998; PWS RCAC 1996a, 1996b, 1997c, 1998, 1999a, 1999b, 2001a, 2002a). In 2001, the PWS RCAC embarked on a collaborative project to deploy a new radar system for marine ice detection in Prince William Sound, located on an island near the Columbia Glacier. This marine ice detection radar system became operational in 2002 through a multilateral collaboration including contributions from the PWS RCAC, the U.S. Coast Guard, the U.S. Army, NOAA, ADEC, the regional oil industry, OSRI, and a local community college. In addition to its ice detection function, the radar site serves as a research and development platform for new technologies designed to enhance marine ice detection (Busenberg 2008; OSRI 2003; PWS RCAC 1999a, 2000a, 2000b, 2001a, 2001b, 2002a, 2002b, 2003a, 2003b). In Cook Inlet, the CIRCAC designed and deployed a network of cameras to allow the

NOAA National Weather Service to monitor ice conditions in the waters of Cook Inlet (CIRCAC 2010).

Significant progress in preparedness for marine and terminal firefighting has also occurred in the marine oil trade of Alaska since 1989. Oil tanker and oil terminal fires can cause massive damages and trigger marine oil spills. Fires therefore present significant hazards in the Alaskan oil trade. Since 1989, significant advances have been made in the training and planning for marine and terminal firefighting in Prince William Sound and Cook Inlet. In 1992, the PWS RCAC collaborated with the U.S. Coast Guard, the oil industry, and a local community to form a fire protection task force in the Sound. The work of the task force led to oil terminal fire protection exercises (Busenberg 2008; PWS RCAC 1993c, 2002a). In 1996, the PWS RCAC funded a study of marine fire response in Prince William Sound that recommended the development of a program to train land-based firefighters in marine firefighting (Busenberg 2008; PWS RCAC 1996a, 1997a, 2003d). The PWS RCAC subsequently collaborated with the oil industry, the U.S. Coast Guard, state and local governments, and a local community college to sponsor a series of marine firefighting training symposia in Prince William Sound to train land-based firefighters in the strategies needed to effectively combat a major fire aboard oil tankers and other marine vessels (Busenberg 2008; PWS RCAC 1997b, 1997c, 1999a, 2000b, 2003d, 2005c). Advances in marine fire protection have also occurred in Cook Inlet. In 2005, the CIRCAC produced marine firefighting plans for the Cook Inlet region developed in collaboration with the U.S. Coast Guard and local firefighters. These marine firefighting plans were subsequently integrated into a vessel fire drill in Cook Inlet (Busenberg 2008; CIRCAC 2003, 2004, 2005).

Significant advances have occurred in the design and use of tug escort vessels to accompany oil tankers in Alaska since 1989. Tug escort vessels can assist in the prevention of oil spills by providing additional propulsion and steering for oil tankers that encounter navigational or mechanical problems. Some tug vessels are also equipped with marine firefighting and oil spill response capabilities. In the aftermath of the *Exxon Valdez* disaster, the state of Alaska in 1989 issued an emergency order requiring that two tug vessels escort each outgoing oil tanker throughout the passage in Prince William Sound (Busenberg 2008). In 1994, federal regulations issued under the authority of OPA 90 required that two tug vessels escort each single-hull tanker laden with oil throughout the passage in Prince William Sound. This federal dual-escort requirement only applied to single-hull tankers, but in practice dual escorts were used for double-hull tankers in the Sound as well. The PWS RCAC nevertheless raised concerns that the escort system might be reduced or eliminated over time, since the federal tanker escort regulations only applied to the single-hull tankers that were being progressively removed from the marine oil trade of Prince William Sound under the requirements of OPA 90. The PWS RCAC proposed a reform to extend the federal requirement for dual escorts to include laden double-hull tankers, and the entire Alaska congressional delegation pursued federal legislation to secure that requirement (PWS RCAC 2009a, 2010). In 2010, federal legislation established a mandate for dual escorts to accompany laden double-hull oil tankers in Prince William Sound (PWS RCAC 2011b). By contrast, no regulatory requirement for a tug escort vessel system has been established for Cook Inlet. There was no tug escort vessel system in Cook Inlet until 2007, when one tug escort vessel was deployed in Cook Inlet as a voluntary measure by the oil industry (CIRCAC 2006, 2010; Loy 2006; PWS RCAC 1993d, 1994c, 2005a, 2005c, 2007a, 2009c).

The tug escort system in Prince William Sound has developed over time through a series of research and development efforts (Busenberg 2008). Beginning in 1991, the PWS RCAC began promoting the idea of using tractor tug escort vessels in the marine oil trade of Prince William Sound (as tractor tug vessels possessed advanced maneuvering capabilities beyond those of the conventional tug vessels then used to escort vessels in the Sound). In 1994, the PWS RCAC recommended that the tug escort system in Prince William Sound be reinforced by the use of new tractor tug escort vessels with enhanced maneuvering and propulsion systems. This proposal was initially opposed by the oil industry on the grounds that tractor tug vessels would not outperform conventional tug vessels in the escort mission (Busenberg 2008; PWS RCAC 1994c, 1995a, 1995b, 1995c). Between 1992 and 1997, the PWS RCAC collaborated with the oil industry, U.S. Coast Guard, and ADEC in two major studies that informed the tug vessel debate by assessing navigational risks in Prince William Sound (Busenberg 2008; PWS RCAC 1991a, 1991b, 1994c). These studies led to operational changes in the tug escort vessel system, and also led to the deployment of one additional tug vessel in Prince William Sound (PWS RCAC 1994c, 1995c, 1997a, 1997b, 1997c). The results of these studies did not determine whether tractor tug vessels would systematically outperform conventional tug vessels in the escort mission. However, the tug vessel debate was further informed by additional collaborative studies that examined (1) different tug vessel technologies and (2) the *best available technology* requirement in Alaska state law (which mandated the use of the best available technology for safeguards against marine oil pollution). These additional studies were conducted as collaborative projects involving the PWS RCAC, ADEC, and the oil industry. Information gathered in these additional studies demonstrated that tractor tug vessels could outmaneuver conventional tug vessels (Busenberg 2008; PWS RCAC 1995a, 1995b, 1997a). With support from Alaska governor Tony Knowles, ADEC held that the advanced maneuvering capabilities of tractor tug vessels constituted the best available technology in the escort mission. Between 1999 and 2001, the regional oil industry responded to the best available technology decision by deploying five new tractor tug escort vessels in Prince William Sound. All five of these new tractor tug vessels were equipped with oil spill response equipment and marine firefighting equipment (Alyeska 2003; Busenberg 2008; PWS RCAC 1997b, 1999b, 2000b, 2009c).

The CIRCAC proposed the use of a tug escort vessel in Cook Inlet, and in 2007 one tug vessel was deployed in Cook Inlet following the grounding of an oil tanker in Cook Inlet in the preceding year (CIRCAC 2010; PWS RCAC 2007c). The deployment of this tug vessel in Cook Inlet was a voluntary measure by the oil industry rather than a response to a legal or regulatory requirement. The deployment of this tug escort vessel represented a significant improvement in the safeguards against marine oil pollution in Cook Inlet, given that there was previously no tug escort vessel system in Cook Inlet. However, this one escort vessel cannot offer the redundant protection of the double escort vessel system used in Prince William Sound.

Since 1989, a number of new projects in environmental research and monitoring have emerged in Cook Inlet and Prince William Sound to gather data that will increase understanding of environmental conditions in coastal Alaska. In the event of future marine oil spills in Alaska, these projects could inform efforts in oil spill response and natural resource damage assessment. The foundation of these projects is a set of environmental monitoring programs in coastal Alaska. OPA 90 required that the new advisory councils establish environmental monitoring programs in Prince William Sound and Cook Inlet. The PWS RCAC therefore established an environmental monitoring program for Prince William Sound, and the CIRCAC established an environmental monitoring program for Cook Inlet (Busenberg 2008; CIRCAC 1993, 1996, 1998, 1999; PWS RCAC 1993b, 2000b, 2004b). The *Exxon Valdez* Oil Spill Trustee Council also established a long-term ecosystem monitoring and research program that encompassed Cook Inlet, Prince William Sound, and adjacent areas (Busenberg 2008; EVOSTC 2004; National Research Council 2002; PWS RCAC 2003b).

In addition to these environmental research and monitoring projects, a number of other environmental research projects have been pursued in Cook Inlet and Prince William Sound since 1989. Beginning in 2001, aerial surveys

of coastal habitats in Cook Inlet and Prince William Sound have been conducted through projects involving the CIRCAC, PWS RCAC, Trustee Council, research institutes, the U.S. National Park Service, Kenai Peninsula
Borough (a local government), and NOAA (Busenberg 2008; CIRCAC 2001,
2002, 2003, 2004; PWS RCAC 2004a). The CIRCAC has collaborated with
ADEC, the U.S. Environmental Protection Agency, and NOAA in research
vessel expeditions to collect data on environmental conditions in Cook Inlet
and the Gulf of Alaska (Busenberg 2008; CIRCAC 2002, 2004). OSRI has
collaborated with the oil industry in periodic vessel expeditions to gather data
on environmental conditions in Prince William Sound. The PWS Science
Center, OSRI, and NOAA have also collaborated in studies of environmental
conditions in Prince William Sound and the Gulf of Alaska through the
deployment of ocean monitoring equipment (Busenberg 2008; CIRCAC
2005; OSRI 2003; PWS RCAC 2005c; PWS Science Center 2004).

Because the oil spill response system in the marine oil trade of Alaska
proved inadequate for the demands of the *Exxon Valdez* oil spill, OPA 90 and
a new Alaska state law enacted after the *Exxon Valdez* disaster created new
oil spill response requirements for the marine oil trade of Alaska (Beaver,
Butler, and Myster 1994). This new legal framework led to the large-scale
reinforcement of the oil spill response systems in the marine oil trades of both
Prince William Sound and Cook Inlet (Alyeska 1987, 2003; Busenberg 2008;
PWS RCAC 2005b; Unocal 1987). In addition to building up large new
stockpiles of oil spill response equipment in both Prince William Sound and
Cook Inlet, the oil industry has also established contracts and training for the
incorporation of local fishing vessels into oil spill response operations in
Prince William Sound. The oil industry has further established plans for the
use of response equipment transported from other regions to Alaska, plans
which would be used in the event that a catastrophic oil spill overwhelmed
the response equipment available in Alaska (Alyeska 2003; PWS RCAC
1993a, 2001a, 2009c). Periodic oil spill response exercises are now held in
both Prince William Sound and Cook Inlet, and these exercises are monitored
by the advisory councils (CIRCAC 2005; PWS RCAC 2009c).

A number of research and development projects related to oil spill
response have been pursued in various projects involving the advisory councils, the PWS Science Center, OSRI, ADEC, NOAA, the oil industry, and
state and local governments. These studies include efforts to map areas in
coastal Alaska exposed to the risk of oil spills, to develop specialized oil spill
response plans that reflect varying conditions found along the coast of Alaska,
to model potential oil spill trajectories in coastal Alaska, and to study
approaches to oil spill response in marine ice conditions (CIRCAC 1991,

1992, 1994, 1995, 1996, 1999, 2002, 2003, 2005; OSRI 2003; PWS RCAC 2005c; PWS Science Center 2004). These research efforts could be used to help guide the deployment of oil spill response assets in the event of an oil spill in the Inlet or the Sound (CIRCAC 1998, 1999, 2000, 2002, 2003; PWS RCAC 1996a, 2003b, 2004a).

The *Exxon Valdez* Oil Spill Trustee Council aimed to protect wildlife populations in the *Exxon Valdez* oil spill region by protecting the threatened habitats of those populations (EVOSTC 2004; Hunt 2009). The Trustee Council invested substantial funds from the *Exxon Valdez* civil settlement to add new conservation units in the *Exxon Valdez* oil spill region. The Trustee Council pursued a range of land conservation efforts in the *Exxon Valdez* spill region in collaboration with regional landowners, land management agencies, and nonprofit conservation organizations. These Trustee Council efforts resulted in the conservation of approximately 647,000 acres in the *Exxon Valdez* spill region by 2009 (EVOSTC 2009). The conservation actions of the Trustee Council included substantial purchases of lands that had been previously transferred to Alaska Native corporations under the institutional arrangements established during the Alaska pipeline punctuation (EVOSTC 2004; Hunt 2009).

As shown in this section, the *Exxon Valdez* oil spill punctuation established a new legal framework and an assortment of new organizations that progressively advanced environmental protection in the marine oil trade of Alaska and the general region of the *Exxon Valdez* spill. The subsequent equilibrium is notable for the high degree of multilateral collaboration between both new and preexisting organizations for the purpose of improving environmental protection in the marine oil trade of Alaska and the *Exxon Valdez* spill region. As shown in the next section, this punctuation also had major policy impacts on the management of the marine oil trades of the United States and the world.

National and International Impacts of the *Exxon Valdez* Oil Spill Punctuation

The *Exxon Valdez* oil spill punctuation created new standards for oil spill prevention and response across the entire maritime jurisdiction of the United States. The implementation of the tanker design standards of OPA 90 led to the progressive replacement of single-hull oil tankers with double-hull oil tankers in the marine oil trade of the United States, and the new oil spill planning provisions of OPA 90 also strengthened the role of the federal government in oil spill response efforts (Randle 2012). A sustained decline in the volume of oil spilled from vessels in American coastal waters occurred in the years following the approval of OPA 90 (Ramseur 2010). Yet while the *Exxon Valdez* oil spill equilibrium led to significant progress in environmental safeguards used in the oil tanker trade, that equilibrium did not lead to comparable progress in the environmental safeguards used in offshore oil drilling. The *Exxon Valdez* disaster was the focusing event and primary driving force leading to the enactment of OPA 90, but this oil tanker disaster led Congress to focus OPA 90 on oil transportation and storage rather than on offshore drilling (Randle 2012, 2). In 2010, the massive oil spill from the offshore drilling platform *Deepwater Horizon* in the Gulf of Mexico demonstrated the continuing risks to the environment posed by marine oil pollution in U.S. waters. The *Deepwater Horizon* disaster also demonstrated the continuing difficulties of marine oil spill response on large scales. The *Deepwater Horizon* oil spill became the largest marine oil spill in U.S. history, replacing the previous record set by the *Exxon Valdez* spill; both disasters clearly demonstrated that marine oil spill prevention was a far more effective method of environmental protection than marine oil spill response (Burger 1997; Freudenburg and Gramling 2010; Randle 2012).

The *Exxon Valdez* oil spill punctuation led to a successful international effort to establish a protective-hull design standard for oil tankers across the world. In 1990, the United States proposed that double hulls be required on tankers worldwide (National Research Council 1998, 26). Previous proposals by the United States for an international requirement for protective hulls on oil tankers had not been adopted, as discussed earlier in this chapter. However, the *Exxon Valdez* oil spill punctuation changed the dynamics of the international dialogue concerning oil tanker hull design by establishing a unilateral U.S. double-hull standard for oil tankers; by 2015, no oil tanker could legally operate without a double hull within U.S. maritime jurisdiction under OPA 90 (National Research Council 1998; Randle 2012). The double-hull oil tanker design mandate of OPA 90 represented a unilateral policy reform with global consequences for the oil tanker trade. With the United States importing vast amounts of oil from other nations through the tanker trade, the double-hull oil tanker design mandate in OPA 90 created global market pressure for the shipping industry to order double-hull oil tankers (Alaska Oil Spill Commission 1990). In 1992, new regulations under the International Convention for the Prevention of Pollution from Ships (MARPOL 73/78) established a worldwide requirement for a gradual transition to the use of oil tankers with double hulls or other types of protective designs (DeSombre

2006; Eyres 2007; National Research Council 1998, vi, 26–33; Tan 2006). The insistence of the United States on the use of the double-hull oil tanker design in U.S. maritime jurisdiction effectively precluded the adoption of alternatives to the double-hull oil tanker design across the world, as it was not commercially viable to build alternative designs that could not operate within U.S. maritime jurisdiction (Eyres 2007; Tan 2006). The MARPOL schedule for the transition to double-hull oil tankers was subsequently accelerated to make double hulls the general design standard for oil tankers worldwide in 2015 (Eyres 2007; Tan 2006). In essence, the OPA 90 policy for the progressive transition to the use of double-hull oil tankers in U.S. maritime jurisdiction became the model for a parallel global policy. The *Exxon Valdez* oil spill punctuation therefore triggered international policy reforms that established the double-hull design as the new standard for oil tankers worldwide.

Summary

The *Exxon Valdez* disaster acted as a focusing event that triggered a critical period of policy reform with enduring consequences for the marine oil trades of Alaska, the United States, and the world. Prior to the *Exxon Valdez* disaster, oil and shipping interests had opposed stronger environmental regulations in the marine oil trade with enough success to prevent a major realignment of U.S. marine oil spill policy. The strong public and political response to the *Exxon Valdez* disaster created a new political opportunity structure that greatly favored the efforts of environmental interests to establish policy reforms to reduce the risk of marine oil pollution. Environmental groups participated heavily in the congressional hearings concerning oil spill reform during the *Exxon Valdez* oil spill punctuation, and the reforms that emerged from this punctuation were heavily weighted toward environmental interests rather than oil and shipping industry interests (Kurtz 2004). The *Exxon Valdez* oil spill punctuation led to enduring policy consequences at the regional, state, national, and international levels. The most important institutional arrangement established in this punctuation was the Oil Pollution Act of 1990, which created new organizations that were designed to enhance the environmental safety of the marine oil trade of Alaska at the regional level, allowed coastal states to continue imposing their own marine oil pollution liability and response requirements within their jurisdictions, strengthened the federal planning and liability standards for marine oil pollution nationwide, mandated double hulls on all oil tankers operating within U.S. maritime jurisdiction by 2015, and created the impetus for an international reform imposing a protective-hull design standard on oil tankers worldwide. The policy image supporting these institutional arrangements reflected a concerted effort to reduce the risk of marine oil pollution in U.S. waters through the strengthening of standards for oil spill prevention, planning, response, and liability across the maritime jurisdiction of the United States.

The *Exxon Valdez* oil spill equilibrium has been marked by the progressive improvement of environmental safeguards in the marine oil trade of Alaska. However, oil production and transportation activities remain inherently hazardous to the environment in Alaska. As shown in chapter 5, efforts to establish new oil development projects in the northern Alaska region have created a continuing national conflict over the potential for these projects to harm the wilderness and wildlife of the American Arctic.

Oil, Wilderness, and Alaska

The Enduring Conflict

T HE THREE CRITICAL PERIODS OF REFORM examined in this book together established enduring policy realignments governing oil development and nature conservation in Alaska. However, these policy realignments did not end the national conflict over oil and wilderness in Alaska. This chapter examines two particularly important political controversies concerning oil and wilderness in the northern Alaska region. For the purposes of this book, the terms *northern Alaska region* and *American Arctic* are used interchangeably to refer to a region containing both Alaska lands north of the Arctic Circle and offshore Arctic areas along the coast of Alaska that are under U.S. jurisdiction. The first controversy involves conflicting efforts for oil development and nature conservation in the Arctic National Wildlife Refuge in northern Alaska. The second controversy involves conflicting efforts for offshore oil development and environmental protection in the Arctic waters along the northern coastline of Alaska. The policy dynamics of these continuing conflicts over oil and wilderness represent the intersecting legacies of the Alaska pipeline punctuation and the Alaska lands conservation punctuation. The Alaska pipeline punctuation created an equilibrium in which the oil industry built up the infrastructures necessary to support large-scale oil development in the northern Alaska region; these infrastructures have positioned the oil industry to seek further expansion of oil and gas development in both onshore and offshore areas of the American Arctic. The Alaska pipeline punctuation also created conditions that facilitated oil development rather than gas development in the northern Alaska region. The Trans-Alaska Pipeline System only transports oil, and no comparable transportation system for natural gas has yet been built in northern Alaska despite the vast reserves of natural gas found on the North Slope of Alaska. Furthermore, the risk of oil spills is the central environmental concern surrounding fossil fuel development in the northern Alaska region. This chapter therefore focuses on oil as the key natural resource currently driving the conflict between fossil fuel projects and environmental protection in the northern Alaska region. The Alaska lands conservation punctuation created an equilibrium in which a large area of the Arctic coastal plain of Alaska was protected as a wildlife refuge, yet also designated as a site to be studied for fossil fuel development. The ongoing conflict between oil development interests and environmental interests in the northern Alaska region therefore reflects the incremental dynamics of two policy equilibria that were established decades ago.

The future of oil and wilderness in Alaska is the subject of longstanding and unresolved debates involving the state of Alaska, the oil industry, the federal government, environmental groups, and Alaska Natives. The state of Alaska supports new fossil fuel development in Alaska due to the critical economic importance of oil in the state and the limitations of developed oil fields in Alaska. The state budget of Alaska is heavily dependent on revenues from oil activities and is vulnerable to declining oil production. The developed oil fields in Alaska will not allow sustainable oil production over the long term. Indeed, oil production from the Prudhoe Bay oil field is declining. The state of Alaska is therefore seeking new energy resources and associated revenues (Busenberg 2011; McBeath et al. 2008; McBeath and Morehouse 1994). For its part, the oil industry has long pursued the expansion of oil development in the northern Alaska region in the face of sustained opposition from environmental groups. Concerns over plans for expanded oil development in the northern Alaska region have also emerged from within the federal government and Alaska Native communities. Environmentalists and federal agencies with environmental protection missions have voiced concerns over the potential environmental impacts of expanded oil development in the northern Alaska region, while Alaska Natives have voiced concerns over the potential threats posed by oil development to subsistence activities that remain vital to some rural communities in Alaska. However, the federal government and some Alaska Native communities also stand to benefit from royalties and taxes on expanded fossil fuel development activities in Alaska and its surrounding waters. Therefore, the federal government and Alaska Natives have expressed a mixture of support and opposition for oil development in the northern Alaska region. The resulting policy dynamics have produced a pattern of incremental policy change. The federal government has long blocked oil exploration and development in the Arctic National Wildlife Refuge but is cautiously proceeding with plans to explore oil resources in the regions offshore of northern Alaska (Reiss 2012).

The debate over oil and environmental protection in the Arctic National Wildlife Refuge is examined next, followed by a discussion of the parallel debate concerning oil and environmental protection in the regions offshore of northern Alaska.

The Arctic National Wildlife Refuge

The Arctic National Wildlife Refuge (Arctic refuge) in northern Alaska has been the focus of a longstanding policy conflict between development interests seeking to open part of the refuge to oil development and environmentalists seeking to preserve the refuge in a wild state. The Arctic refuge is the largest land unit of the National Wildlife Refuge System (Woods 2003). The Arctic refuge includes a coastal plain located near the North Slope oil industry complex that has long been considered a promising prospect for oil development (H.R. Rep. No. 95-1045 Part I 1978, 368). The Arctic refuge was assembled into its current form under the Alaska National Interest Lands Conservation Act (ANILCA) in a compromise between competing environmental and development interests. ANILCA protected much of the Arctic refuge as a wilderness area but also designated a section of the refuge's coastal plain as an area to be studied for its fossil fuel development potential (Layzer 2011). Under ANILCA, congressional approval is needed for fossil fuel development to proceed in the coastal plain of the Arctic refuge. If federal approval is granted and significant oil production proves feasible in the coastal plain of the Arctic refuge, pipelines can be built to connect oil drilling sites in the Arctic refuge to the Trans-Alaska Pipeline System. The oil industry complex of the North Slope of Alaska would thereby be enlarged to encompass the coastal plain of the Arctic refuge. However, the federal approval necessary for oil development in the coastal plain of the Arctic refuge has not been granted in the decades of political conflict over the fate of the Arctic refuge and its abundant wildlife (Ross 2000).

The development of the Arctic National Wildlife Refuge occurred through a series of policy reforms long promoted by the environmental movement.

Conservation in northern Alaska was a particular concern of Bob Marshall, who was one of the founders and funders of the environmental organization The Wilderness Society (Sutter 2002). In 1938, Bob Marshall proposed that northern Alaska be protected largely as a wilderness (Kaye 2006, 1–2; Nash 2001, 288). The Wilderness Society would subsequently join with other environmental organizations to promote wilderness preservation in northern Alaska. In the period 1952–60, a coalition of activists from the environmental movement and their allies in the federal government mounted a concerted campaign to establish a large wildlife reserve in northern Alaska (Kaye 2006).

In 1950, the U.S. National Park Service launched a survey of Alaska's recreation potential (Kaye 2006, 14–18; Williss 2005, 9). George Collins and Lowell Sumner of the National Park Service were both involved in the Alaska recreation survey, and they collaborated with the environmental movement to promote the idea of large-scale conservation in northern Alaska (Kaye 2006, 28, 65). In 1952, Collins and Sumner proposed the creation of a large conservation area in northern Alaska and northern Canada for the protection of wilderness and wildlife, including wildlife that migrated across the Arctic border of Alaska and Canada (Kaye 2006, 35–39). In 1953, Collins and Sumner published their proposal for a protected area in the North American Arctic in the *Sierra Club Bulletin* (Collins and Sumner 1953). The national environmental movement and its allies in the federal government mounted a sustained and widely publicized national effort to establish a protected area in northern Alaska. In 1957, the directors of the National Park Service, Fish and Wildlife Service, and Bureau of Land Management met together with representatives of the environmental organizations The Wilderness Society, Sierra Club, and National Parks Association to coordinate efforts to establish a protected area in Arctic Alaska. Those present at this meeting concluded that a wildlife range proposal was the best course of action. A wildlife range in Arctic Alaska could accommodate hunting, mining, and the subsistence activities of the Alaska Natives of the region; as all of these activities would likely be precluded in a national park, the wildlife range proposal might face less opposition than a proposal for a national park. A wildlife range was also less likely to lead to the road-building that was common in the tourism-oriented national park system (H.R. Rep. No. 95-1045 Part I 1978, 100; Kaye 2006, 116–18). The wildlife range proposal aimed to establish a protected area in Arctic Alaska that would maintain existing subsistence uses and limit recreation-oriented development, so that the area protected would likely remain largely as it was. The intended conservation design of the Arctic range thereby balanced multiple land uses (nature conservation, sport hunting, subsistence, and mining) in a manner that set a precedent for the subsequent conservation design of ANILCA. Environmentalists and their allies in the federal government pursued the effort to establish an Arctic range alongside efforts to establish two smaller wildlife conservation areas (Kuskokwim and Izembek Bay) elsewhere in Alaska (Kaye 2006, 127).

In 1957, Interior Secretary Fred Seaton proposed the establishment of a wildlife range in Arctic Alaska, and also proposed to open part of northern Alaska to development by modifying a preexisting policy (Public Land Order 82) that had withdrawn much of northern Alaska for military use during the Second World War (Kaye 2006, 133). Seaton thereby proposed a new set of land use policies for northern Alaska that would balance development and conservation interests. The proposal to establish an Arctic range was supported by environmental and sporting organizations, but opposed by development interests (particularly mining interests). Opposition to the Arctic range proposal delayed action on that proposal (Kaye 2006, 136). Alaska became a state in the meantime, and the new Alaska state government swiftly took a position against the Arctic range proposal. A resolution passed in the first session of the Alaska state legislature opposed the establishment of an Arctic wildlife conservation area on the grounds that such an action would discourage development in northern Alaska (Kaye 2006, 157). Federal legislation to establish the Arctic range passed in the House of Representatives in 1960, but the Alaska delegation in the Senate successfully acted to prevent a companion bill from reaching a vote in the full Senate. Alaska governor William Egan then requested that the area proposed for a federal Arctic range instead be transferred to the state of Alaska (Kaye 2006, 161–62, 202–4). Faced with state opposition and congressional inaction, Interior Secretary Seaton acted decisively in favor of the conservationists by establishing the Arctic National Wildlife Range in 1960 through Public Land Order 2214 (Kaye 2006, 205; Norris et al. 1999; U.S. Department of Interior 1987). In 1960, Secretary Seaton further announced the establishment of the Izembek National Wildlife Range and Kuskokwim National Wildlife Range in Alaska (Kaye 2006, 206). Secretary Seaton also revoked the longstanding federal withdrawal of 20 million acres of lands in northern Alaska (Kaye 2006, 205; Norris et al. 1999). These 20 million acres of northern Alaska (including the Prudhoe Bay region) were thereby made available for state land selections and development (Kaye 2006, 205–6). Another large area of northern Alaska was kept under federal control as a national petroleum reserve (McBeath et al. 2008). The Alaska governor and Alaska congressional delegation proceeded to lobby Interior Secretary Stewart Udall (who replaced Seaton in that position) to reverse the executive establishment of the Arctic National Wildlife Range, but to no avail (Kaye 2006, 208–9). The Alaska congressional delegation succeeded in blocking appropriations for the management of the Arctic range until 1969, when the Fish and Wildlife Service was finally able to hire its first manager for the range (Kaye 2006, 210).

In 1980, ANILCA more than doubled the size of the Arctic range, renamed it the Arctic National Wildlife Refuge, and designated a large part of the Arctic refuge as a wilderness (U.S. Department of the Interior 1987; U.S. Fish and Wildlife Service 1988). In 1983, the state of Alaska relinquished to the federal government 1.3 million acres of state land selections in an area largely surrounded by the Arctic National Wildlife Refuge. ANILCA allowed the federal government to incorporate donated lands into the conservation units in Alaska, and in 1983 the federal government used this authority to incorporate 971,000 acres of the lands relinquished by the state into the Arctic National Wildlife Refuge (U.S. Fish and Wildlife Service 1988). In 1988, Congress added an additional 325,000 acres of the relinquished lands to the Arctic National Wildlife Refuge, thereby making the Arctic refuge the largest land unit in the National Wildlife Refuge System; the Arctic refuge now protects more than 19 million acres of northern Alaska (U.S. Fish and Wildlife Service 2008b).

ANILCA designated approximately 1.5 million acres of the coastal plain of the Arctic refuge as an area to be considered for oil and gas development, while also prohibiting oil and gas development in the Arctic refuge pending further congressional action on the subject. In essence, Congress reserved the authority to grant approval for oil and gas development in the Arctic National Wildlife Refuge (Layzer 2011, 113; U.S. Department of the Interior 1987; U.S. Fish and Wildlife Service 1988). ANILCA also required the Interior secretary to conduct studies to describe the fish and wildlife—and the potential for oil and gas development—of the coastal plain of the Arctic refuge (Ross 2000; U.S. Department of the Interior 1987, 2–3, 7–8). A national conflict subsequently emerged over proposals to open the Arctic National Wildlife Refuge for oil and gas development, with the state of Alaska and the oil industry promoting fossil fuel development in the refuge against sustained opposition from the national environmental movement. A 1987 federal study of the coastal plain of the Arctic National Wildlife Refuge found that this area contained "the most outstanding petroleum exploration target in the onshore United States" (U.S. Department of the Interior 1987, 7). In the same year, Interior Secretary Donald Hodel recommended the approval of oil and gas development in the Arctic refuge, noting the proximity of the coastal plain of the Arctic refuge to the North Slope oil industry complex and the Trans-Alaska pipeline. The North Slope oil industry complex could provide a well-developed base of operations for oil development and exploration in the Arctic refuge, and oil from the Arctic refuge could be transported to market through the Trans-Alaska Pipeline System. Oil development in the Arctic refuge therefore offered the potential to extend the use of the Trans-Alaska Pipeline System (U.S. Department of the Interior 1987, 185–88).

Environmentalists claimed that fossil fuel development in the coastal plain of the Arctic National Wildlife Refuge would damage the refuge's natural qualities and cause harm to its populations of caribou, polar bear, and other species (Layzer 2011, 119–20). Environmentalists proceeded to mount a sustained national campaign to protect the coastal plain of the Arctic National Wildlife Refuge (Bosso 2005; Guber and Bosso 2007; H.R. Rep. No. 95-1045 Part I 1978, 384; Layzer 2011, 134). The environmentalist position on the Arctic refuge was reinforced by a number of international wildlife treaties signed by the United States that were designed to protect migratory species found in the Arctic National Wildlife Refuge, as some of these species could be harmed by fossil fuel development in the refuge (Docherty 2001; U.S. Department of the Interior 1987, 5). Indeed, one of the purposes of the Arctic National Wildlife Refuge stated in ANILCA was "to fulfill the international treaty obligations of the United States with respect to fish and wildlife and their habitats" (94 Stat. 2390). A prominent example of an international migratory wildlife population found in the Arctic National Wildlife Refuge was the Porcupine caribou herd, named after the Porcupine River that the herd often crossed during migrations (Woods 2003, 32). A 1979 House of Representatives report on Alaska lands legislation noted the migration of the Porcupine caribou herd across the border between Canada and Alaska, and also noted that the Canadian government had withdrawn an area of 9 million acres with the intent of establishing a wilderness park to protect the range of the caribou herd in Canada (H.R. Rep. No. 96-97 Part I 1979, 182, 485). In the United States, the protection of the U.S. range of the Porcupine caribou herd was one of the purposes of the 1980 expansion of the Arctic refuge under ANILCA (H.R. Rep. No. 96-97 Part I 1979, 485). The United States and Canada later signed the 1987 Agreement on the Conservation of the Porcupine Caribou Herd with the intent of protecting this migratory wildlife population (Docherty 2001). Plans for fossil fuel development in the Arctic refuge carried the potential to harm the Porcupine caribou herd and thereby raised conflicts with the protective ANILCA language governing the refuge and international treaty obligations concerning the Porcupine caribou herd (Docherty 2001; Layzer 2011, 136; U.S. Department of the Interior 1987;

Woods 2003). The environmentalist position on the Arctic refuge was also supported by a National Research Council report finding substantial environmental impacts associated with the North Slope oil industry complex, thereby indicating that fossil fuel development in the Arctic National Wildlife Refuge would likely cause substantial environmental impacts (National Research Council 2003; Ross 2000). The environmentalist position on the Arctic refuge was further supported by the considerable scientific value of the Arctic refuge, as the refuge provided a uniquely undisturbed land region within the American Arctic that could be compared with more developed areas of the Arctic to assess the environmental impacts of development in the Arctic (H.R. Rep. No. 96-97 Part I 1979, 487, 585).

Repeated attempts to achieve congressional approval for oil development in the Arctic National Wildlife Refuge have been met with a long series of setbacks. The 1989 *Exxon Valdez* disaster was credited with impairing the political momentum for oil development in the Arctic National Wildlife Refuge, and repeated efforts to rebuild that momentum have met with political defeat (Layzer 2011). In the absence of federal approval for fossil fuel development within the Arctic National Wildlife Refuge, the entire refuge has remained a wild area (Kaye 2006, 219). As the twentieth century came to a close, the 1952 proposal by Collins and Sumner for an international protected area in the North American Arctic was effectively realized when Canada established the Ivvavik and Vuntut national parks that connected to the Arctic National Wildlife Refuge across the international border (Kaye 2006; Kopas 2007; Rennicke 1995). U.S. environmentalists continue to urge the federal government to protect the coastal plain of the Arctic National Wildlife Refuge from development by designating that coastal plain as a wilderness.

While the Arctic National Wildlife Refuge remains off limits to oil development to date, some areas offshore of northern Alaska are under consideration for fossil fuel development in the face of sustained opposition from environmentalists. The continuing conflict over oil development plans in areas offshore of Arctic Alaska is examined in the next section.

Offshore Drilling, Climate Change, and Environmental Protection in the American Arctic

The Arctic Ocean is one of the great oil and gas frontiers of the twenty-first century (Reiss 2012). However, the offshore exploration and development of Arctic oil and gas poses major environmental risks for this ecologically fragile region. Oil development in the Arctic Ocean is particularly threatening to the environment of the region due to the risk of high wildlife casualties in the event of a marine oil spill combined with the challenges of oil spill response in the extreme conditions of the Arctic (Coen 2012; Oil Spill Commission Action 2012). This section examines the ongoing policy conflict over oil development and environmental protection in areas offshore of northern Alaska, beginning with an examination of the complex policy history by which the federal government and the state of Alaska came to claim jurisdiction over natural resources found offshore of Alaska.

Offshore drilling projects and efforts to protect the marine environment in the American Arctic are based on U.S. claims of maritime jurisdiction extending from the coast of Alaska. These maritime jurisdictions developed over time as the result of a complex series of policy reforms that successively extended state and federal jurisdictions seaward from the U.S. coastline. These reforms have provided the basis for vast state and federal claims over the offshore areas surrounding Alaska, with the large scale of these maritime claims reflecting the extensive coastline and continental shelf areas of Alaska. The coastline of Alaska extends across more than six thousand miles, and a large continental shelf area (a submerged area covered in relatively shallow waters) extends seaward from the Alaskan coastline (Mason et al. 1997). Valuable natural resources such as oil, gas, and fish are often found in continental shelf areas. The state of Alaska now claims jurisdiction over natural resources in a maritime zone generally extending three geographical miles seaward from the coastline of Alaska; beyond that boundary of state maritime jurisdiction, the federal government now claims jurisdiction over natural resources in a maritime zone generally extending 200 nautical miles seaward from the Alaska coastline (Wilder 1998). State maritime jurisdictions are designated using the geographical mile as a measure of distance, while federal maritime jurisdictions are designated using the nautical mile as a measure of distance; the two measures of distance differ only slightly. The distance of 3 geographical miles that marks the general limit of state maritime claims extending seaward from the coastline of Alaska is equivalent to a distance of 3.45 statute or land miles, and the distance of 200 nautical miles that marks the general limit of federal maritime claims extending seaward from the coastline of Alaska is equivalent to a distance of 230 statute or land miles (National Institute of Standards and Technology 2011). These jurisdictional claims represent the culmination of a series of national reforms designed to progressively expand federal and state jurisdictions seaward in attempts to secure control of marine resources such as offshore oil fields and fisheries.

A series of national policy reforms in the period 1945–83 had the overall effect of greatly expanding U.S. control of natural resources found offshore of the nation's coastline. In 1945, President Harry Truman signed a proclamation asserting federal jurisdiction over the natural resources of the seabed and subsoil of the U.S. continental shelf, thereby creating a federal claim to offshore oil and gas reserves contiguous to the U.S. coast (Truman 1945a). A subsequent proclamation signed by Truman in 1945 asserted federal jurisdiction to regulate and control fishing activities in the high seas contiguous to the U.S. coast (Truman 1945b). These executive proclamations by Truman established federal jurisdiction over oil, gas, fish, and other natural resources in large offshore areas around Alaska. Subsequent federal legislation codified the federal claim to much of the U.S. continental shelf, but also codified state maritime jurisdictions off the U.S. coastline. The 1953 Submerged Lands Act established state jurisdictions over submerged lands and their resources out to three geographical miles seaward from the coastlines of most coastal states; these state maritime jurisdictions could be further limited by international maritime boundaries (67 Stat. 31). The 1953 Outer Continental Shelf Lands Act codified the claim of the federal government to the U.S. continental shelf beyond state waters (Wilder 1998). The Alaska Statehood Act confirmed the applicability of the Submerged Lands Act to the new state of Alaska (72 Stat. 343). Upon Alaska statehood in 1959, the Submerged Lands Act granted the state of Alaska claim to the submerged lands and natural resources in a maritime area generally extending out to three geographical miles seaward from the Alaska coast. The divided system of state and federal maritime jurisdictions also gave the federal government claim to the submerged lands and natural resources in the remaining continental shelf areas along the coast of Alaska. Both state and federal maritime jurisdictions are further limited by international maritime boundaries around Alaska (H.R. Rep. No. 96–97 Part I 1979, 582).

In the 1976 Fishery Conservation and Management Act, the United States claimed exclusive control of fisheries extending from the seaward limits of state maritime jurisdictions out to two hundred nautical miles seaward from the U.S. coast (Bean and Rowland 1997; Christie and Hildreth 2007, 364–68; Layzer 2011). A subsequent treaty established an international framework through which coastal nations could claim large natural resource management zones at sea. The 1982 United Nations Convention on the Law of the Sea allowed each signatory nation exclusive control of natural resources in an exclusive economic zone extending two hundred nautical miles seaward from its coastline (although the extent of this zone could be truncated by international maritime boundaries). The Law of the Sea convention also allowed claims on the continental shelf to be extended past two hundred nautical miles seaward from coastlines. The Law of the Sea convention thereby attempted to create a worldwide system of maritime zoning. The United States did not join the Law of the Sea convention on the basis of objections to the deep seabed mining provisions of that convention. Nevertheless, in 1983, President Ronald Reagan unilaterally proclaimed an exclusive economic zone extending two hundred nautical miles seaward of the U.S. coastline based on a claim of such zones constituting customary international law (Reagan 1983a, 1983b). The United States thereby claimed the largest exclusive economic zone in the world (Christie and Hildreth 2007; Sohn et al. 2010, 247–49; Wilder 1998, 81). The Fishery Conservation and Management Act was later amended to incorporate the exclusive economic zone designation, replacing the earlier fishery zone designation in the Act (Christie and Hildreth 2007, 364–68).

The long coastline of Alaska provided the baseline for a large exclusive economic zone claim for the United States in the Arctic Ocean, although this claim was truncated by adjoining maritime boundaries with Russia and Canada. The continental shelf and exclusive economic zone claims of America established the basis for U.S. jurisdiction over natural resource extraction and management activities far off the coast of Alaska. Indeed, the coastline of Alaska constitutes the only basis for U.S. maritime jurisdiction in the Arctic Ocean.

The establishment of state and federal maritime jurisdictions off the coast of Alaska has created the basis for large areas offshore of northern Alaska to be leased for oil and gas development. However, the prospect of offshore oil development in the northern Alaska region raises a number of particularly acute environmental concerns. Offshore oil drilling facilities are capable of generating enormous oil spills, as demonstrated by the 2010 *Deepwater Horizon* disaster in the Gulf of Mexico (Freudenburg and Gramling 2010; Randle 2012). Oil production in offshore areas of the American Arctic would also require a new transportation system to move the oil to the market, with attendant risks to the environment. Oil produced in offshore areas of the American Arctic could be transported by a combination of submerged and overland pipelines connecting to the Trans-Alaska Pipeline System, or it could be transported by pipelines connecting to a marine oil terminal in northern Alaska that would load the oil onto icebreaking oil tankers (Reiss 2012). The history of the Trans-Alaska Pipeline System indicates that these new oil transportation systems would create new risks of oil spills and environmental damage. Therefore, plans for offshore oil development in the American Arctic have created a new field of conflict between environmental and development interests.

The dynamics of the conflict between environmental and development interests in the American Arctic can be illustrated by examining the debate over polar bears and their habitat in the Arctic sea ice—a habitat that overlaps with offshore oil and gas leases around the coastline of northern Alaska. Due to the vast reach of polar bear habitat in the American Arctic, the protection of that habitat holds broad potential to provide for the conservation of marine and coastal ecosystems in the Arctic.

The conflict between offshore drilling and polar bear conservation rose to prominence as a confluence of events made fossil fuel development in Arctic waters increasingly feasible and economically desirable. As shown in chapter 2, the Arctic test voyages of the icebreaking oil tanker *Manhattan* demonstrated the difficulties of using oil tankers in the sea ice of the Arctic Ocean. However, the *Manhattan* did not represent a technological dead end for Arctic oil transportation by sea. While the technical challenges of marine oil transportation in Arctic waters were perceived as excessively difficult and risky at the time of the Alaska pipeline punctuation, subsequent advances have allowed icebreaking oil tankers and offshore oil loading terminals to come into use in the sea ice conditions of the Arctic waters of Russia (Eyres 2007). Furthermore, the Arctic sea ice appears to be progressively melting due to climate change (Arctic Climate Impact Assessment 2005). Sea ice is a dynamic feature of the Arctic that expands in extent across the Arctic Ocean in winter and retreats in summer. The annual average extent of the Arctic sea ice has diminished considerably since the time of the Alaska pipeline punctuation (Arctic Climate Impact Assessment 2005). The rapid decline of the Arctic sea ice generally reduces the barriers posed by sea ice to offshore drilling and marine transportation in the Arctic Ocean (Arctic Climate Impact Assessment 2005; Coen 2012; Reiss 2012).

The decline of the Arctic sea ice poses a grave threat to the future of the polar bears found in the Arctic region (U.S. Fish and Wildlife Service 2008a). The polar bear is a marine mammal that makes use of the sea ice as a platform for walking and hunting far across the frozen surface of the Arctic Ocean. The sea ice is the primary habitat of the polar bear, and the progressive reduction of Arctic sea ice due to climate change therefore threatens the polar bear by eroding its primary habitat. The diminishing sea ice of the Arctic is also eroding the habitat of several species of ice seals that are adapted to life in sea ice conditions; these ice seals are the primary prey of the polar bear (Arctic Climate Impact Assessment 2005; Center for Biological Diversity 2005, 2008b; U.S. Fish and Wildlife Service 2008a, 2010).

Polar bear conservation efforts in Alaska began with a focus on the sport hunting of polar bears, then progressed to focus on the potential impacts of oil and gas development and declining sea ice cover on polar bear populations. While polar bear hunting in Alaska was traditionally pursued on a limited scale by Alaska Natives, a rise in sport hunting using airplanes increased pressure on polar bear populations in Alaska in the twentieth century. The sport hunting of polar bears in Alaska ended in 1972, when the state of Alaska banned all sport hunting of polar bears and the federal Marine Mammal Protection Act of 1972 (Pub. L. 92-522) placed a national moratorium on the taking of marine mammals (including polar bears), albeit with some exceptions allowing for limited takings such as traditional subsistence hunting (Bean and Rowland 1997, 490; Ross 2000, 10). In 1973, an international treaty to protect polar bear populations (the Agreement on the Conservation of Polar Bears) was concluded by the United States, Canada, the Soviet Union, Denmark, and Norway (Bean and Rowland 1997, 490–91).

As plans for offshore oil and gas development in Arctic Alaska progressed, environmental groups pursued a series of legal actions aimed at securing additional protections for the polar bears in that region. In 2005, the environmental organization Center for Biological Diversity petitioned the U.S. Fish and Wildlife Service to list the polar bear as a threatened species under the 1973 Endangered Species Act (Pub. L. 93-205), based on the claim that the reduction in Arctic sea ice cover due to climate change threatened the polar bear with extinction in the wild (Center for Biological Diversity 2005). The environmental organizations Natural Resources Defense Council and Greenpeace joined the Center for Biological Diversity in this petition and in a lawsuit intended to compel a response to the petition from the Fish and Wildlife Service. The Fish and Wildlife Service agreed to respond to the polar bear petition through a settlement with the environmental organizations involved in this lawsuit, and in 2006, the Fish and Wildlife Service proposed to list the polar bear as a threatened species under the Endangered Species Act, thereby creating a basis for federal actions to protect the polar bear. However, the final polar bear listing decision under the Endangered Species Act was delayed as the oil industry pursued offshore oil exploration programs in the American Arctic. In 2008, the Center for Biological Diversity, Natural Resources Defense Council, and Greenpeace filed another lawsuit against the Fish and Wildlife Service claiming that the agency had missed the deadline for a final listing decision for the polar bear. In 2008, a federal judge ordered the Fish and Wildlife Service to issue a final listing decision for the polar bear by May 15, 2008. One day before this court-ordered deadline, Interior Secretary Dirk Kempthorne announced the final listing of the polar bear as a threatened species under the Endangered Species Act due to the threat of polar bear habitat loss as the Arctic sea ice cover receded. Yet Kempthorne also announced that the threatened species listing for the polar bear would not be used to influence U.S. climate change policy and further announced that natural resources development in the American Arctic would proceed (Center for Biological Diversity 2012; U.S. Department of the Interior 2008; U.S. Fish and Wildlife Service 2008a).

The listing of the polar bear as a threatened species was followed by a series of legal actions not only by environmental groups seeking additional protections for the polar bear but also by the state of Alaska for opposing purposes. The conflict over the polar bear had revealed that the Endangered Species Act could be used as a means to interfere with oil industry exploration and development activities in the areas offshore of northern Alaska. The state of Alaska proceeded to sustain its longstanding commitment to the promotion of natural resource development by filing a lawsuit seeking to remove the threatened species listing for the polar bear, but to no avail. By contrast, legal actions by environmentalists seeking to strengthen protections for the polar bear resulted in a settlement proposing an enormous critical habitat designation for the polar bear in the offshore areas around Alaska.

Under the Endangered Species Act, critical habitat refers to areas designated as essential for the conservation of species listed as threatened or endangered under the act (Bean and Rowland 1997, 202). Federal agencies are prohibited from authorizing activities that are likely to destroy or adversely modify critical habitat (U.S. Fish and Wildlife Service 2010). In 2009, the Fish and Wildlife Service reached a partial settlement with environmental groups under which the agency proposed to designate a large critical habitat for the polar bear. In 2010, the Fish and Wildlife Service designated a polar bear critical habitat of approximately 120 million acres along the coast of Alaska, consisting primarily of offshore regions where the sea ice habitat of the polar bear was found (Center for Biological Diversity 2012; U.S. Fish and Wildlife Service 2010). The critical habitat designated for the polar bear encompassed substantial areas offshore of Alaska that had been leased for oil and gas development. The overlap between the polar bear critical habitat designation and offshore drilling leases created the potential for conflict between drilling plans and measures to protect the polar bear, as federal authorization was needed for drilling in these overlapping areas. The oil and gas industry, the state of Alaska, and Alaska Native groups responded with legal actions seeking to prevent the enforcement of the critical habitat designation for the polar bear. In 2013, a federal district court found that the procedures followed by the Fish and Wildlife Service in establishing the polar bear critical habitat rule did not meet the requirements of either the Endangered Species Act or the 1946 Administrative Procedure Act. The court decision stated that the critical habitat designation for the polar bear "presents a disconnect between the twin goals of protecting a cherished resource and allowing for growth and much needed economic development. The current designation went too far and was too extensive" (*Alaska Oil & Gas Association v. Salazar* 2013, 63). The court vacated the polar bear critical habitat rule and remanded that rule to the Fish and Wildlife Service for revision.

The polar bear dispute illustrates the sweeping potential of federal environmental laws such as the Endangered Species Act to impede oil exploration and development activities in the American Arctic. The polar bear dispute is not an isolated case. Indeed, environmental groups and Alaska Natives have pursued multiple venues of appeal against offshore oil drilling that might harm the environment of the American Arctic. For example, environmental groups have petitioned the federal government to protect not only the polar bear population but also a number of other marine mammal populations (including seals, whales, and walruses) with habitats in the Arctic Ocean (Center for Biological Diversity 2005, 2008a, 2008b; Center for Biological Diversity and Marine Biodiversity Protection Center 2000; Reiss 2012, 68). In the period 2007–9, plans for offshore oil drilling in the American Arctic were halted by environmental concerns raised in legal challenges by environmental groups and Alaska Native groups (Reiss 2012, 11). Subsequent actions by the federal government halted plans for offshore oil drilling in the American Arctic in the period 2010–11. In 2010, the federal government temporarily suspended offshore oil drilling in the American Arctic in response to the *Deepwater Horizon* oil-spill disaster (Reiss 2012, 21). After this drilling suspension was lifted, the U.S. Environmental Protection Agency delayed the approval of an air quality permit needed for offshore oil drilling in the American Arctic (Reiss 2012, 185, 213–14, 253–54). The combined pressure from environmentalists, Alaska Natives, and federal agencies with environmental protection missions has thereby caused years of delay in the offshore drilling program for the American Arctic. In 2012, the oil industry received federal authorization for preliminary offshore drilling activities in the American Arctic, but a series of mishaps associated with the drilling program have led to additional delays and government reviews that leave the future of the drilling program in question. This continuing policy conflict over oil drilling and environmental protection in areas offshore of northern Alaska has so far yielded an unquiet compromise in which the federal government has both sustained a commitment to continue Arctic oil exploration and taken a series of actions designed to protect the environment of the American Arctic (including an expanded Arctic research program). In response to evolving federal requirements, environmental concerns, and lessons learned from the *Deepwater Horizon* disaster, the oil industry has created an elaborate environmental protection plan for the American Arctic drilling program. This industry plan calls for a collection of vessels and equipment to provide safeguards against marine oil pollution in the offshore drilling program in the American Arctic—including two drilling ships that can drill relief wells in the event of a blowout, an upgraded blowout preventer, an undersea oil spill containment device, and support vessels with oil spill response capabilities (Reiss 2012, 172–76). However, a series of mishaps with the vessels and equipment used in the offshore drilling program for the American Arctic have served to underscore the inherent risks of Arctic offshore drilling.

Plans for oil exploration and development in the American Arctic carry the prospect of severe risks to the environment of that region because conditions in the Arctic Ocean pose extraordinary challenges for the environmental safety of oil industry operations. The problem of marine oil spill response in the Arctic Ocean is a particular concern. Despite all industry precautions, the remote location, limited infrastructure, severe climate conditions, and sea ice of the Arctic Ocean could make it difficult or impossible to effectively operate oil spill response vessels and equipment in that region (Coen 2012, 167; Oil Spill Commission Action 2012). Overall, the Arctic Ocean poses some of the greatest challenges for marine oil spill response in the world. Offshore oil development efforts in Arctic Alaska therefore pose massive threats to the environment of that region, and call into question the commitment of the nation to protect the marine and coastal ecosystems found around the coastline of northern Alaska.

Summary

The continuing drive for oil development in the American Arctic suggests that the three critical periods of policy reform examined in this book together set the stage for enduring policy conflicts over oil and wilderness in the northern Alaska region. Environmental concerns have played a pivotal role in slowing the expansion of oil drilling in the northern Alaska region, as demonstrated both by the delay in offshore drilling plans for the American Arctic and by the stalemate over drilling proposals for the Arctic National Wildlife Refuge. The sweeping implications of climate change in the Arctic have only intensified the conflict over economic and environmental values in Alaska.

Climate change is now simultaneously opening new opportunities for natural resource exploitation in the American Arctic with the retreat of the Arctic sea ice, and raising concerns that habitat changes caused by a rapidly shifting climate could pose major threats to the wildlife and ecosystems of Alaska (Reiss 2012).

6

Conclusion

THIS BOOK HAS EXAMINED three major cases of policy reform that match the patterns of policy change predicted by the punctuated equilibrium theory (Baumgartner and Jones 2009). In each case, an equilibrium period of incremental policy change was punctuated by a critical period of major policy reform. Each critical period was marked by heightened media and congressional attention supportive of reform and by the authorization of new institutional arrangements and policy images that established a major realignment of policy. The institutional arrangements and policy images established in each critical period have endured for decades in subsequent equilibrium periods, with enduring policy consequences. Each of the three critical periods of reform established institutional legacies that subsequently defined three parallel equilibrium periods. The Alaska Native Claims Settlement Act and the Trans-Alaska Pipeline Authorization Act are the institutional legacies of the Alaska pipeline punctuation, together defining a continuing equilibrium in which oil development has come to dominate the economy of the state of Alaska and in which Alaska Natives have taken possession of large areas of Alaska (McBeath et al. 2008). The Trans-Alaska Pipeline System and the North Slope oil industry complex are the enduring infrastructural legacies of the Alaska pipeline punctuation. The Alaska National Interest Lands Conservation Act is the institutional legacy of the Alaska lands conservation punctuation, a law defining a continuing equilibrium in which Alaska holds most of the total land areas in the national park, national wildlife refuge, and national wilderness preservation systems of

America (Andrus and Freemuth 2006). The institutional legacies of the *Exxon Valdez* oil spill punctuation include the Oil Pollution Act of 1990, new state laws, and new organizations designed to reduce the risk of marine oil pollution. These institutional legacies together define a third continuing equilibrium in which the safeguards against marine oil pollution in Alaska and the United States have been progressively reinforced. Therefore, the three critical periods examined in this study have created enduring institutional legacies (in the form of laws and organizations) and enduring policy consequences (in the form of oil infrastructures, protected natural areas, and oil pollution safeguards). A timeline of key events and punctuations examined in this book is provided in figure 6.1.

While the evidence presented in this book supports the central predictions of the punctuated equilibrium theory of policy change, that same evidence does not provide comparable support for the central predictions of three alternative models of the policy process that differ significantly from the punctuated equilibrium theory in their portrayals of policy change over time. Each of these alternative models (policy incrementalism, issue-attention cycle, and policy erosion or reversal) describes important possible features of the policy process. However, the punctuated equilibrium theory more accurately depicts the policy reforms examined in this book than do these alternative models. These three alternative models of the policy process are examined next in the context of the evidence presented in this study.

Policy incrementalism is a model of the policy process predicting a pattern of incremental policy change over time. Lindblom (1959) proposed that the limited capacity of administrators to foresee the consequences of policy change would lead to a pattern of policy incrementalism. For administrators, the potential consequences of policy alternatives would be most easily understood when those alternatives consisted of incremental adjustments to existing policies. These incremental policy alternatives could be readily compared with existing policies. In an effort to simplify the decision process and avoid major errors, administrators would focus their attention selectively on these incremental policy alternatives. A succession of incremental policy changes would allow administrators to progress gradually toward their goals without risking serious mistakes that might result from abrupt, radical policy shifts. Incremental adjustments to policy would therefore characterize the policy process, and the policy process would not contain fundamental realignments in policy (Lindblom 1959). This study finds three examples of fundamental policy realignments that do not match the predictions of the policy incrementalism model. The policy incrementalism model does not provide an accurate description of the critical periods of policy reform that authorized the building of the Alaska pipeline, the creation of a new system of conservation for Alaska

Figure 6.1. Timeline of Key Events and Punctuations Examined in This Book

Year	Event
1867	Alaska purchased by the United States from Russia.
1959	Alaska statehood.
1968	Alaska pipeline punctuation begins with the announcement of the Prudhoe Bay oil field discovery in the North Slope region of Alaska.
1969	Trans-Alaska Pipeline System proposed by oil industry.
1970	Alaska lands conservation punctuation begins with legislative proposals for conservation planning in Alaska.
1971	Alaska Native Claims Settlement Act serves to clear the route of the Alaska pipeline and establishes a lands conservation planning process for Alaska.
1972	Nixon administration withdraws large areas of Alaska for conservation planning purposes as authorized by the Alaska Native Claims Settlement Act.
1973	Trans-Alaska Pipeline Authorization Act ends the Alaska pipeline punctuation.
1977	Trans-Alaska Pipeline System begins shipments of oil, Alaska lands conservation bill H.R. 39 introduced in the U.S. House of Representatives.
1978	Carter administration protects large areas of Alaska for conservation purposes.
1980	Alaska National Interest Lands Conservation Act ends the Alaska lands conservation punctuation.
1989	*Exxon Valdez* disaster in Alaska acts as a focusing event that begins the *Exxon Valdez* oil spill punctuation.
1990	Oil Pollution Act of 1990 ends the *Exxon Valdez* oil spill punctuation.

lands, and the founding of a new institutional framework for the management of marine oil pollution in Alaska and America. These realigning reforms represented major departures from preexisting policies rather than incremental adjustments to those established policies. While incremental policy change is a defining feature of the equilibria examined in this study, a model of policy incrementalism cannot account for the punctuations examined in this study.

Downs (1972) proposed an issue-attention cycle model of the policy process in which a rise in public attention to an issue in a given policy domain would swiftly fade away even if the issue remained largely unresolved. While a rise in public attention to an issue would place pressure on political leaders to respond, political leaders could respond to this pressure through symbolic reassurances rather than fundamental reforms (Edelman 1964). The frequent cycling of public attention among different policy issues would subsequently remove the impetus for reform in the policy domain (Downs 1972). The issue-attention cycle model therefore focused on the possibility that major shifts in public attention would not be met by major shifts in policy. Contrary to the predictions of the issue-attention cycle model, the major reforms examined in this study provide three examples in which major shifts in public attention (as measured by media attention) were met with major shifts in policy in the form of the three punctuations examined in this study, each of which was accompanied by a rise in media attention to the issues in question. While media attention cycled away from those issues over time, the political actions taken in each critical period were primarily characterized by fundamental and substantive policy reforms rather than symbolic reassurances. The issue-attention cycle model cannot account for these major reforms.

A model of policy erosion or reversal proposed by Patashnik (2008) predicts the possibility of major policy reforms being diminished or revoked over time as the result of sustained attacks by opponents of the reforms. This model allows for the possibility of fundamental reform, but also recognizes that opposition to reform can outlast the enactment of reform. Therefore, fundamental reforms can be attacked and diminished during their implementation. This book provides little evidence to support this model of policy erosion or reversal. Indeed, all three of the major policy reforms examined in this book have survived intact for decades. The evidence presented in this study therefore does not support a model of policy erosion or reversal, although it is certainly possible that the major reforms examined in this book could be diminished or revoked in the future.

The major policy reforms examined in this study therefore display a common pattern of policy change that differs from the patterns predicted by the models of policy erosion or reversal, issue-attention cycle, and policy incrementalism. While each of these three alternative models of policy change clearly describes significant dimensions of the policy process in general, none of them suffices to accurately depict the development of policies for oil and wilderness in Alaska. Instead, the pattern of punctuated equilibria is the dominant feature of the major reforms examined in this book.

While this study finds a recurring pattern of punctuated equilibria across three periods of fundamental policy reform, this study also finds significant variations in the policy dynamics of each reform period and the policy realignment it established. A comparison of the Alaska lands conservation punctuation with the *Exxon Valdez* oil spill punctuation is particularly useful in revealing the many differences in the processes by which punctuations can be triggered and driven. The *Exxon Valdez* oil spill punctuation was triggered by the *Exxon Valdez* disaster, which acted as a focusing event that precipitated an intense wave of media and political attention supportive of reforming the environmental management of the marine oil trade in America. By contrast, the Alaska lands conservation punctuation was not directly triggered by a focusing event, but instead was instigated by the opportunistic intervention of the national environmental movement and their allies in the policy processes leading to the settlement of the Alaska Native land claims. These two punctuations differed dramatically in their congressional dynamics. The *Exxon Valdez* oil spill punctuation was characterized by the rapid emergence of strong congressional support for marine oil spill reform in a manner that favored the interests of the environmental movement in this policy domain, leading to a relatively swift punctuation that lasted less than two years. By contrast, the Alaska lands conservation punctuation was characterized by a period of congressional inaction followed by a sustained conflict between the causes of Alaska conservation and Alaska development in Congress, drawing out the Alaska lands conservation punctuation for a period of ten years (Williss 2005). The executive branch played differing roles in the two punctuations as well. The *Exxon Valdez* oil spill punctuation was dominated by legislative actions. By contrast, executive actions played a pivotal role in the Alaska lands conservation punctuation. Unilateral conservation actions taken by the Carter administration in 1978 and 1980 played a key role in impelling the resolution of the Alaska lands conservation debate in Congress. Finally, the two punctuations differed in the extent of their compromises between competing interests. The Alaska lands conservation punctuation was marked by a complex compromise among competing interests that balanced demands for conservation with numerous demands for other uses of Alaska lands, while the *Exxon Valdez* oil spill punctuation yielded a reform that was weighted in favor of environmental interests demanding stronger regulation of the marine oil trade. The two critical periods of reform therefore differed in their origins, in the extent of the legislative controversy involved in the reforms, in the extent of executive actions involved in the reforms, in their duration, and in the extent of compromise among competing interests achieved by the reforms.

Both of these punctuations differed in turn from the Alaska pipeline punctuation. When compared to the *Exxon Valdez* oil spill punctuation, the Alaska pipeline punctuation was triggered by a focusing event in the form of a newly discovered economic opportunity (the great Prudhoe Bay oil field) rather than by a crisis-driven focusing event in the form of an environmental disaster (the *Exxon Valdez* oil spill). The Alaska pipeline punctuation was character-ized by a compromise that balanced demands for oil development, Alaska Native land claims, and conservation planning in Alaska; as noted above, the *Exxon Valdez* oil spill punctuation produced no comparable compromise among competing environmental and industry interests. The Alaska pipeline punctuation was resolved over a period of five years through the approval of two national laws and was therefore a longer and more complex period of reform than the *Exxon Valdez* oil spill punctuation. However, the Alaska pipe-line punctuation was concluded in approximately half the time required to conclude the Alaska lands conservation punctuation. And while the Alaska lands conservation punctuation was notable for unilateral executive actions that impelled a conclusion to the prolonged congressional debate on the subject of Alaska lands conservation, the Alaska pipeline punctuation was driven by congressional actions accompanied by sustained executive persua-sion rather than executive unilateralism.

The punctuations examined in this book therefore differed widely in their dynamics. While the patterns of policy change in all three punctuations exam-ined in this book closely matched the central predictions of the punctuated equilibrium theory, each of these punctuations followed a distinct path of policy reform. These punctuations also had two important features in com-mon beyond the patterns predicted by the theory. First, each critical period of reform was informed by extensive prior studies of the issues relevant to reform; in each critical period, the policy processes leading to reform were deliberate and well-informed. Second, each critical period of reform was driven by national interests that extended beyond the borders of Alaska. The issues of domestic oil production, lands conservation, and oil pollution man-agement dealt with in these punctuations were of national importance and were dealt with accordingly at the national level. The state of Alaska was unable to occupy a commanding position in these reform efforts driven by national interests, even though these reform efforts fundamentally reshaped modern Alaska. Instead, the state of Alaska was compelled to play a role similar to the roles played by interest groups (such as environmental interests or industry interests) in the national policy processes that led to these reforms.

The three critical periods of reform examined in this book established three equilibrium periods with an essential common feature predicted by the punctuated equilibrium theory: the durability over time of the institutions and ideas associated with each equilibrium period. However, these equilibria differed widely in their dynamics and policy consequences for Alaska. For the purpose of clarity, each equilibrium period is given a title matching the title of the punctuation that established it. The Alaska pipeline punctuation established the Alaska pipeline equilibrium, the Alaska lands conservation punctuation established the Alaska lands conservation equilibrium, and the *Exxon Valdez* oil spill punctuation established the *Exxon Valdez* oil spill equilibrium. The Alaska pipeline equilibrium focused on the construction and operation of oil infrastructures in Alaska, creating an enduring dynamic of oil development as the dominant form of economic activity in Alaska. The Alaska lands conservation equilibrium focused on the preservation of large conservation units in Alaska, creating an enduring dynamic of nature preservation, outdoor recreation, sport hunting, and subsistence as leading land uses across large areas of Alaska. These two equilibria had vastly different consequences for the landscape areas they affected in Alaska. The Alaska pipeline equilibrium led to a large-scale transformation of part of the Alaskan landscape through the building and operation of one of the largest industrial complexes in the world in northern Alaska and a very long pipeline extending across large areas of the state. By contrast, the Alaska lands conservation equilibrium essentially preserved vast areas of the Alaskan landscape as wilderness. The *Exxon Valdez* oil spill equilibrium differed in turn from the other equilibria examined in this book. The *Exxon Valdez* oil spill equilibrium did not lead to the building of new infrastructures across landscape scales, and the additions to conservation units made during this equilibrium were relatively limited in scale when compared to the lands protected by ANILCA. The most striking feature of the *Exxon Valdez* oil spill equilibrium is its enduring dynamic of progressive enhancements to environmental safeguards in the marine oil trades of Alaska, the United States, and the world.

The three cases of major policy reform examined in this book are thereby characterized by a general pattern of punctuated equilibria, yet also characterized by wide variations in the dynamics and enduring consequences of reform. The complex and varying dynamics of these three punctuations serve to illustrate a limitation of the punctuated equilibrium theory. While the punctuated equilibrium theory provides a clear conceptual map for exploring the dynamics of past policy reforms and their subsequent consequences, the theory does not provide a systematic basis for predicting future policy reforms (True, Jones, and Baumgartner 2007). This point is illustrated by the three critical periods of reform examined in this study, each of which was fundamentally shaped by events that could not have been reliably predicted. The Alaska pipeline punctuation resulted from the discovery of the Prudhoe Bay oil field. The Alaska lands conservation punctuation was fundamentally shaped by the election of President Carter and his relentless pursuit of conservation in Alaska. The *Exxon Valdez* oil spill punctuation was fundamentally driven by the experience of the *Exxon Valdez* disaster. In essence, the three punctuations examined in this book were driven by unpredictable events. It therefore follows that the punctuated equilibrium theory cannot be used to precisely predict future policy reforms in Alaska, because any such reforms might depend on future events (such as another major oil field discovery in Alaska) that cannot be reliably predicted.

The three punctuations examined in this book effectively represent a sequence of interacting policy reforms. The conservation planning process authorized by the Alaska pipeline punctuation set the stage for the subsequent Alaska lands conservation punctuation. The oil infrastructures authorized by the Alaska pipeline punctuation eventually led to the *Exxon Valdez* disaster, which in turn triggered the *Exxon Valdez* oil spill punctuation. Therefore, the Alaska pipeline punctuation set the stage for the other two punctuations examined in this book. Furthermore, this study finds two cases in which critical periods of reform at the national level led to additional reforms with major international consequences. The Arctic tanker tests that occurred during the Alaska pipeline punctuation triggered a cascade of policy reforms through which Canada claimed jurisdiction over a vast reach of the Arctic Ocean, an international repercussion that was neither proposed nor supported by the United States. The *Exxon Valdez* oil spill punctuation led to the adoption of a protective hull design standard for oil tankers worldwide, an international repercussion that was proposed and supported by the United States. These two cases provide important examples of national policy reforms with major international repercussions.

The findings of this book have implications for further empirical research on the punctuated equilibrium theory. In light of the widely varying dynamics and consequences of the reform efforts examined in this book, it is evident that further empirical research applying the theory would benefit from the use of research methods that allow the detailed examination of the processes and consequences of reform. Important variations in the dynamics and consequences of different reform efforts can only be fully understood by studying the details of reform. Case study and historical research methods are therefore indispensable in providing the detailed knowledge necessary to fully understand policy reforms. This study also demonstrates important interactions between different reform efforts. The possibility of interacting reforms should therefore be considered in further empirical research applying the punctuated equilibrium theory, including the possibility of critical periods of reform at the national level that lead to major international repercussions.

The common theme that connects all of the major policy reforms examined in this book is the enduring conflict between environmental interests supporting nature conservation in Alaska and various other interests pursuing natural resource development in Alaska. Oil is by far the most important of the natural resources extracted in Alaska to date, and the conflict between oil development and wilderness values in Alaska is at the center of the policy processes examined in this book. Contained within the persistent conflict over oil and wilderness in Alaska are two competing national visions of Alaska. One vision focuses on oil in Alaska as the driving force of the Alaska state economy and a major element of American oil production. Another vision focuses on Alaska as the state with the greatest areas of land in the nation that still possess the essential qualities of wilderness. The reforms and continuing policy conflicts examined in this book reflect the efforts of the nation to balance the values of oil development and environmental protection in Alaska. Those efforts have caused environmental damage to large areas of the land and coastal waters of Alaska, but they have also established the largest system of land conservation found in any state in the nation. In light of the extraordinary commitment made by the nation to land and wildlife protection in Alaska, continuing efforts to balance oil and wilderness values in Alaska will involve difficult questions of natural resource development and environmental protection that will be of pivotal importance for the future of American nature conservation.

NOTES

Chapter 2. The Trans-Alaska Pipeline System

This chapter is an adaptation and expansion of George J. Busenberg, "The Policy Dynamics of the Trans-Alaska Pipeline System," *Review of Policy Research* 28:5 (2011). © The Policy Studies Organization.

Chapter 4. The *Exxon Valdez* Disaster and the Oil Pollution Act of 1990

The descriptions of safeguards against marine oil pollution in Alaska found in this chapter expand on work that appeared in George J. Busenberg, "Managing the Hazard of Marine Oil Pollution in Alaska," *Review of Policy Research* 25:3 (2008).

REFERENCES

Alaska Oil & Gas Association v. Salazar 2013. 2013 U.S. Dist. LEXIS 10559, 43 Envtl. L. Rep. 20013 (D. Alaska 2013).

Alaska Oil Spill Commission. 1990. *Spill: The Wreck of the Exxon Valdez: Implications for Safe Transportation of Oil.* 5 vols. Juneau, AK: State of Alaska.

Alcock, Tammy M. 1992. "'Ecology Tankers' and the Oil Pollution Act of 1990: A History of Efforts to Require Double Hulls on Oil Tankers." *Ecology Law Quarterly* 19: 97–145.

Allin, Craig W. 2008. *The Politics of Wilderness Preservation.* Fairbanks, AK: University of Alaska Press. First published 1982 by Greenwood Press.

Alyeska (Alyeska Pipeline Service Company). n.d. *Memorandum re Trans-Canada Alternate Route.* Anchorage, AK: Alyeska Pipeline Service Company.

—————. 1987. *Oil Spill Contingency Plan, Prince William Sound.* Anchorage, AK: Alyeska Pipeline Service Company.

—————. 2003. *Trans Alaska Pipeline System Facts.* Anchorage, AK: Alyeska Pipeline Service Company.

—————. 2009. *Facts: Trans Alaska Pipeline System.* Anchorage, AK: Alyeska Pipeline Service Company.

Andrews, Richard N. L. 2006. *Managing the Environment, Managing Ourselves: A History of American Environmental Policy.* 2nd ed. New Haven, CT: Yale University Press.

Andrus, Cecil D., and John C. Freemuth. 2006. "President Carter's Coup: An Insider's View of the 1978 Alaska Monument Designations." In *The Antiquities Act: A Century of American Archaeology, Historic Preservation, and Nature Conservation,* edited by David Harmon, Francis P. McManamon, and Dwight T. Pitcaithley, 93–105. Tucson: University of Arizona Press.

Arctic Climate Impact Assessment. 2005. *Arctic Climate Impact Assessment.* Cambridge, UK: Cambridge University Press.

Ashenmiller, Joshua. 2006. "The Alaska Oil Pipeline as an Internal Improvement, 1969–1973." *Pacific Historical Review* 75(3): 461–89.

Banner, Stuart. 2007. *Possessing the Pacific: Land, Settlers, and Indigenous People from Australia to Alaska.* Cambridge, MA: Harvard University Press.

Baumgartner, Frank R., and Bryan D. Jones. 2009. *Agendas and Instability in American Politics.* 2nd ed. Chicago: University of Chicago Press.

Bean, Michael J., and Melanie J. Rowland. 1997. *The Evolution of National Wildlife Law.* 3rd ed. Westport, CT: Praeger.

Beaver, James E., James N. Butler III, and Susan E. Myster. 1994. "Stormy Seas? Analysis of New Oil Pollution Laws in the West Coast States." *Santa Clara Law Review* 34(3): 791–839.

Berry, Mary Clay. 1975. *The Alaska Pipeline: The Politics of Oil and Native Land Claims.* Bloomington: Indiana University Press.

Birkland, Thomas A. 1997. *After Disaster: Agenda Setting, Public Policy, and Focusing Events.* Washington, DC: Georgetown University Press.

Birkland, Thomas A., and Regina G. Lawrence. 2001. "The *Exxon Valdez* and Alaska in the American Imagination." In *American Disasters,* edited by Steven Biel, 382–402. New York: New York University Press.

Black, Lydia T. 2004. *Russians in Alaska, 1732–1867.* Fairbanks: University of Alaska Press.

Bockstoce, John R. 2009. *Furs and Frontiers in the Far North: The Contest among Native and Foreign Nations for the Bering Strait Fur Trade.* New Haven, CT: Yale University Press.

Bohlen, Janet Trowbridge. 1993. *For the Wild Places: Profiles in Conservation.* Washington, DC: Island Press.

Bosso, Christopher J. 2005. *Environment, Inc.: From Grassroots to Beltway.* Lawrence: University Press of Kansas.

Boyce, John R., and Mats A. N. Nilsson. 1999. "Interest Group Competition and the Alaska Native Land Claims Settlement Act." *Natural Resources Journal* 39(4): 755–98.

Brew, David A. 1974. *Environmental Impact Analysis: The Example of the Proposed Trans-Alaska Pipeline.* Geological Survey Circular 695. Reston, VA: U.S. Geological Survey.

Brinkley, Douglas. 2009. *The Wilderness Warrior: Theodore Roosevelt and the Crusade for America.* New York: HarperCollins Publishers.

———. 2011. *The Quiet World: Saving Alaska's Wilderness Kingdom, 1879–1960.* New York: HarperCollins Publishers.

Burger, Joanna. 1997. *Oil Spills.* New Brunswick, NJ: Rutgers University Press.

Busenberg, George J. 2008. "Managing the Hazard of Marine Oil Pollution in Alaska." *Review of Policy Research* 25(3): 203–18.

———. 2011. "The Policy Dynamics of the Trans-Alaska Pipeline System." *Review of Policy Research* 28(5): 401–22.

Bush, George. "White House Fact Sheet on Federal Assistance for the Alaskan Oil Spill Cleanup." April 7, 1989. In *Public Papers of the Presidents of the United States: George Bush, 1989*, bk. 1. Washington, DC: U.S. Government Printing Office.

―――. "Statement on Signing the Oil Pollution Act of 1990." August 18, 1990. In *Public Papers of the Presidents of the United States: George Bush, 1990*, bk. 2. Washington, DC: U.S. Government Printing Office.

Bushell, Sharon, and Stan Jones. 2009. *The Spill: Personal Stories from the Exxon Valdez Disaster.* Kenmore, WA: Epicenter Press.

Cahn, Robert. 1982. *The Fight to Save Wild Alaska.* New York: National Audubon Society.

Carr, Kathleen B. 1992. "Limiting Limitation: *In re The Glacier Bay*." *Tulane Maritime Law Journal* 16(2): 403–9.

Carson, Donald W., and James W. Johnson. 2001. *Mo: The Life and Times of Morris K. Udall.* Tucson: University of Arizona Press.

Carter, Jimmy. 1977. "The Environment Message to the Congress." May 23, 1977. In *Public Papers of the Presidents of the United States: Jimmy Carter, 1977*, bk. 1. Washington, DC: U.S. Government Printing Office.

―――. 1978a. "The State of the Union Annual Message to the Congress." January 19, 1978. In *Public Papers of the Presidents of the United States: Jimmy Carter, 1978*, bk. 1. Washington, DC: U.S. Government Printing Office.

―――. 1978b. "Designation of National Monuments in Alaska Statement by the President." December 1, 1978. In *Public Papers of the Presidents of the United States: Jimmy Carter, 1978*, bk. 2. Washington, DC: U.S. Government Printing Office.

―――. 1979a. "The State of the Union Annual Message to the Congress." January 25, 1979. In *Public Papers of the Presidents of the United States: Jimmy Carter, 1979*, bk. 1. Washington, DC: U.S. Government Printing Office.

―――. 1979b. "Conservationist of the Year Award Remarks on Accepting the Award from the National Wildlife Federation." March 20, 1979. In *Public Papers of the Presidents of the United States: Jimmy Carter, 1979*, bk. 1. Washington, DC: U.S. Government Printing Office.

―――. 1979c. "Environmental Priorities and Programs Message to the Congress." August 2, 1979. In *Public Papers of the Presidents of the United States: Jimmy Carter, 1979*, bk. 2. Washington, DC: U.S. Government Printing Office.

―――. 1980a. "The State of the Union Annual Message to the Congress." January 21, 1980. In *Public Papers of the Presidents of the United States: Jimmy Carter, 1980–81*, bk. 1. Washington, DC: U.S. Government Printing Office.

―――. 1980b. "Alaska Lands Conservation Bill Statement on House Approval of the Legislation." November 13, 1980. In *Public Papers of the Presidents of the United*

States: Jimmy Carter, 1980–81, bk. 3. Washington, DC: U.S. Government Printing Office.

————. 1980c. "Alaska National Interest Lands Conservation Act Remarks on Signing H.R. 39 Into Law." December 2, 1980. In *Public Papers of the Presidents of the United States: Jimmy Carter, 1980–81,* bk. 3. Washington, DC: U.S. Government Printing Office.

Catton, Theodore. 1997. *Inhabited Wilderness: Indians, Eskimos, and National Parks in Alaska.* Albuquerque: University of New Mexico Press.

Center for Biological Diversity. 2005. *Before the Secretary of the Interior: Petition to List the Polar Bear* (Ursus maritimus) *as a Threatened Species under the Endangered Species Act.* Tucson, AZ: Center for Biological Diversity.

————. 2008a. *Before the Secretary of Interior: Petition to List the Pacific Walrus* (Odobenus Rosmaurs Divergens) *as a Threatened or Endangered Species under the Endangered Species Act.* Tucson, AZ: Center for Biological Diversity.

————. 2008b. *Before the Secretary of Commerce: Petition to List Three Seal Species under the Endangered Species Act: Ringed Seal* (Pusa Hispida), *Bearded Seal* (Erignathus Barbatus), *and Spotted Seal* (Phoca Largha). Tucson, AZ: Center for Biological Diversity.

————. 2012. "Polar Bear Action Timeline." Accessed July 9, 2012. www.biological diversity.org/species/mammals/polar_bear/action_timeline.html.

Center for Biological Diversity and Marine Biodiversity Protection Center. 2000. *Before the Secretary of Commerce: Petition to Designate Critical Habitat for the Bering-Chukchi-Beaufort Stock of the Bowhead Whale* (Baleana Mysticetus) *under the Endangered Species Act.* Tucson, AZ: Center for Biological Diversity.

Christie, Donna R., and Richard G. Hildreth. 2007. *Coastal and Ocean Management Law in a Nutshell.* 3rd ed. St. Paul, MN: Thomson/West.

Cicchetti, Charles J. 1972. *Alaskan Oil: Alternative Routes and Markets.* Baltimore, MD: Resources for the Future and Johns Hopkins University Press.

Cicchetti, Charles J., and A. Myrick Freeman III. 1973. "The Trans-Alaska Pipeline: An Economic Analysis of Alternatives." In *Pollution, Resources, and the Environment,* edited by Alain C. Enthoven and A. Myrick Freeman III, 271–84. New York: W. W. Norton.

CIRCAC (Cook Inlet Regional Citizens Advisory Council). 1991. *Annual Report.* Kenai, AK: CIRCAC.

————. 1992. *Annual Report 1992.* Kenai, AK: CIRCAC.

————. 1993. *Annual Report 1993.* Kenai, AK: CIRCAC.

————. 1994. *Annual Report 1994.* Kenai, AK: CIRCAC.

————. 1995. *Annual Report 1995.* Kenai, AK: CIRCAC.

————. 1996. *Annual Report 1996.* Kenai, AK: CIRCAC.

————. 1998. *Annual Report 1998.* Kenai, AK: CIRCAC.

————. 1999. *1999 Annual Report*. Kenai, AK: CIRCAC.

————. 2000. *2000 Annual Report*. Kenai, AK: CIRCAC.

————. 2001. *2001 Annual Report*. Kenai, AK: CIRCAC.

————. 2002. *2002 Annual Report*. Kenai, AK: CIRCAC.

————. 2003. *2003 Annual Report*. Kenai, AK: CIRCAC.

————. 2004. *2004 Annual Report*. Kenai, AK: CIRCAC.

————. 2005. *Annual Report 2005*. Kenai, AK: CIRCAC.

————. 2006. *2006 Annual Report*. Kenai, AK: CIRCAC.

————. 2010. *2010 Annual Report*. Kenai, AK: CIRCAC.

Clarke, Lee B. 1993. "The Disqualification Heuristic: When Do Organizations Misperceive Risk?" *Research in Social Problems and Public Policy* 5: 289–312.

————. 1999. *Mission Improbable: Using Fantasy Documents to Tame Disaster*. Chicago: University of Chicago Press.

Coates, Peter A. 1993. *The Trans-Alaska Pipeline Controversy: Technology, Conservation, and the Frontier*. Fairbanks: University of Alaska Press. First published 1991 by Lehigh University Press and Associated University Presses.

Coen, Ross A. 2012. *Breaking Ice for Arctic Oil: The Epic Voyage of the* SS Manhattan *through the Northwest Passage*. Fairbanks: University of Alaska Press.

Collins, George L., and Lowell Sumner. 1953. "Northeast Arctic: The Last Great Wilderness." *Sierra Club Bulletin* 38: 13–26.

Davis, Charles, ed. 2001. *Western Public Lands and Environmental Politics*. 2nd ed. Boulder, CO: Westview Press.

DeSombre, Elizabeth R. 2006. *Global Environmental Institutions*. New York: Routledge.

Docherty, Bonnie. 2001. "Challenging Boundaries: The Arctic National Wildlife Refuge and International Environmental Law Protection." *New York University Environmental Law Journal* 10(1): 70–116.

Dolin, Eric Jay. 2010. *Fur, Fortune, and Empire: The Epic History of the Fur Trade in America*. New York: W. W. Norton.

Downs, Anthony. 1972. "Up and Down with Ecology—The 'Issue-Attention Cycle.'" *The Public Interest* 28: 38–50.

Duffy, David Cameron, Keith Boggs, Randall H. Hagenstein, Robert Lipkin, and Julie A. Michaelson. 1999. "Landscape Assessment of the Degree of Protection of Alaska's Terrestrial Biodiversity." *Conservation Biology* 13(6): 1332–43.

Durbin, Kathie. 2005. *Tongass: Pulp Politics and the Fight for the Alaska Rain Forest*. 2nd ed. Corvallis: Oregon State University Press.

Duscha, Julius. 1981. "How the Alaska Act Was Won." *The Living Wilderness* 44 (Spring): 4–9.

Edelman, Murray. 1964. *The Symbolic Uses of Politics*. Urbana: University of Illinois Press.

Elliott, Gary E. 1994. *Senator Alan Bible and the Politics of the New West.* Reno: University of Nevada Press.

Elliot-Meisel, Elizabeth. 2009. "Politics, Pride, and Precedent: The United States and Canada in the Northwest Passage." *Ocean Development & International Law* 40: 204–32.

EVOSTC (*Exxon Valdez* Oil Spill Trustee Council). 2004. *Then and Now—A Message of Hope: 15th Anniversary of the Exxon Valdez Oil Spill.* Anchorage, AK: *Exxon Valdez* Oil Spill Trustee Council.

———. 2009. *2009 Status Report.* Anchorage, AK: *Exxon Valdez* Oil Spill Trustee Council.

Eyres, D. J. 2007. *Ship Construction.* 6th ed. Oxford, UK: Butterworth-Heinemann.

Fagan, Brian M. 2004. *The Great Journey: The Peopling of Ancient America.* Updated ed. Gainesville: University Press of Florida.

Federal Field Committee for Development Planning in Alaska. 1968. *Alaska Natives and the Land.* Anchorage, AK: Federal Field Committee for Development Planning in Alaska.

Fineberg, Richard A. 2004. *Trans-Alaska Pipeline System Dismantling, Removal and Restoration (DR&R): Background Report and Recommendations.* Anchorage, AK: PWS RCAC.

Fink, Richard J. 1994. "The National Wildlife Refuges: Theory, Practice, and Prospect." *Harvard Environmental Law Review* 18: 1–135.

Fischman, Robert L. 2003. *The National Wildlife Refuges: Coordinating a Conservation System through Law.* Washington, DC: Island Press.

———. 2005. "The Significance of National Wildlife Refuges in the Development of U.S. Conservation Policy." *Journal of Land Use and Environmental Law* 21: 1–22.

Flippen, J. Brooks. 2000. *Nixon and the Environment.* Albuquerque: University of New Mexico Press.

Freudenburg, William R., and Robert Gramling. 2010. *Blowout in the Gulf: The BP Oil Spill Disaster and the Future of Energy in America.* Cambridge, MA: MIT Press.

GAO (U.S. General Accounting Office). 1993. *Natural Resources Restoration: Use of Exxon Valdez Oil Spill Settlement Funds.* GAO/RCED-93-206BR. Washington, DC: General Accounting Office.

———. 2002. *Alaska's North Slope: Requirements for Restoring Lands After Oil Production Ceases.* GAO-02-357. Washington, DC: General Accounting Office.

Ginsburg, Patty, Scott Sterling, and Sheila Gotteherer. 1993. "The Citizens' Advisory Council as a Means of Mitigating Environmental Impacts of Terminal and Tanker Operations." *Marine Policy* 17(5): 404–11.

Grant, Shelagh D. 2010. *Polar Imperative: A History of Arctic Sovereignty in North America.* Vancouver, Canada: Douglas and McIntyre.

Griffiths, Franklyn, ed. 1987. *Politics of the Northwest Passage*. Kingston, Canada: McGill-Queen's University Press.

Grumbles, Benjamin H., and Joan M. Manley. 1995. "The Oil Pollution Act of 1990: Legislation in the Wake of a Crisis." *Natural Resources & Environment* 10(2): 35–42.

Guber, Deborah Lynn, and Christopher J. Bosso. 2007. "Framing ANWR: Citizens, Consumers, and the Privileged Position of Business." In *Business and Environmental Policy: Corporate Interests in the American Political System*, edited by Michael E. Kraft and Sheldon Kamieniecki, 35–59. Cambridge, MA: MIT Press.

Harmon, David, Francis P. McManamon, and Dwight T. Pitcaithley, eds. 2006. *The Antiquities Act: A Century of American Archaeology, Historic Preservation, and Nature Conservation*. Tucson: University of Arizona Press.

Hartzog, George B., Jr. 1988. *Battling for the National Parks*. Mt. Kisco, NY: Moyer Bell Limited.

Harvey, Mark. 2005. *Wilderness Forever: Howard Zahniser and the Path to the Wilderness Act*. Seattle: University of Washington Press.

Haycox, Stephen W. 2006. *Alaska: An American Colony*. Seattle: University of Washington Press. First published 2002 by University of Washington Press.

Hays, Samuel P. 1999. *Conservation and the Gospel of Efficiency: The Progressive Conservation Movement, 1890–1920*. Pittsburgh: University of Pittsburgh Press. First published 1959 by Harvard University Press.

———. 2007. *Wars in the Woods: The Rise of Ecological Forestry in America*. Pittsburgh: University of Pittsburgh Press.

———. 2009. *The American People and the National Forests: The First Century of the U.S. Forest Service*. Pittsburgh: University of Pittsburgh Press.

H.R. Rep. (U.S. House of Representatives Report) No. 93-414. 1973. *Amending Section 28 of the Mineral Leasing Act of 1920, and Authorizing a Trans-Alaska Oil and Gas Pipeline, and for Other Purposes*. Washington, DC: U.S. Government Printing Office.

H.R. Rep. (U.S. House of Representatives Report) No. 95-5. 1977. *Trans-Alaska Oil Pipeline Safety Features*. Washington, DC: U.S. Government Printing Office.

H.R. Rep. (U.S. House of Representatives Report) No. 95-1045 Part I. 1978. *Alaska National Interest Lands Conservation Act of 1978*. Washington, DC: U.S. Government Printing Office.

H.R. Rep. (U.S. House of Representatives Report) No. 96-97 Part I. 1979. *Alaska National Interest Lands Conservation Act of 1979*. Washington, DC: U.S. Government Printing Office.

H.R. Rep. (U.S. House of Representatives Report) No. 101-242 Part 2. 1989. *Oil Pollution Prevention, Removal, Liability and Compensation Act of 1989*. Washington, DC: U.S. Government Printing Office.

Hunt, Joe. 2009. *Mission without a Map: The Politics and Policies of Restoration following the Exxon Valdez Oil Spill: 1989–2002.* Rev. ed. Anchorage, AK: *Exxon Valdez* Oil Spill Trustee Council.

Hunter, Celia, and Ginny Wood. 1981. "Alaska National Interest Lands." *Alaska Geographic* 8(4): 1–240.

Kaye, Roger. 2006. *Last Great Wilderness: The Campaign to Establish the Arctic National Wildlife Refuge.* Fairbanks: University of Alaska Press.

Keating, Bern, and Dan Guravich. 1970. *The Northwest Passage from the Mathew to the Manhattan: 1497 to 1969.* Chicago, IL: Rand McNally.

Kirkey, Christopher. 1996. "The Arctic Waters Pollution Prevention Initiatives: Canada's Response to an American Challenge." *International Journal of Canadian Studies* 13: 41–59.

———. 1997. "Moving Alaskan Oil to Market: Canadian National Interests and the Trans-Alaska Pipeline, 1968–73." *American Review of Canadian Studies* 27(4): 495–522.

Kopas, Paul S. 2007. *Taking the Air: Ideas and Change in Canada's National Parks.* Vancouver: UBC Press.

Kurtz, Rick S. 2004. "Coastal Oil Pollution: Spills, Crisis, and Policy Change." *Review of Policy Research* 21(2): 201–19.

Layzer, Judith A. 2011. "Oil versus Wilderness in the Arctic National Wildlife Refuge." In *The Environmental Case: Translating Values into Policy*, 3rd ed., 109–39. Washington, DC: CQ Press.

Lindblom, Charles E. 1959. "The Science of 'Muddling Through.'" *Public Administration Review* 19(2): 79–88.

Loy, W. 2006. "On-Duty Tugs Not Present in Cook Inlet." *Anchorage Daily News,* February 3, 2006, p. A1.

Marchetti, Michael P., and Peter B. Moyle. 2010. *Protecting Life on Earth: An Introduction to the Science of Conservation.* Berkeley: University of California Press.

Mason, Owen, William J. Neal, Orrin H. Pilkey, Jane Bullock, Ted Fathauer, Deborah Pilkey, and Douglas Swanston. 1997. *Living with the Coast of Alaska.* Durham, NC: Duke University Press.

Matz, George. 1999. "World Heritage Wilderness: From the Wrangells to Glacier Bay." *Alaska Geographic* 26(2): 1–112.

Maugeri, Leonardo. 2006. *The Age of Oil: The Mythology, History, and Future of the World's Most Controversial Resource.* Westport, CT: Praeger.

McBeath, Gerald A., and Thomas A. Morehouse. 1994. *Alaska Politics and Government.* Lincoln: University of Nebraska Press.

McBeath, Jerry, Matthew Berman, Jonathan Rosenberg, and Mary F. Ehrlander. 2008. *The Political Economy of Oil in Alaska: Multinationals vs. the State.* Boulder, CO: Lynne Rienner Publishers.

Miles, John C. 2009. *Wilderness in National Parks: Playground or Preserve.* Seattle: University of Washington Press.

Millard, Elizabeth R. 1993. "Anatomy of an Oil Spill: The Exxon Valdez and the Oil Pollution Act of 1990." *Seton Hall Legislative Journal* 18(1): 331–69.

Mitchell, Donald Craig. 2001. *Take My Land, Take My Life: The Story of Congress's Historic Settlement of Alaska Native Land Claims, 1960–1971.* Fairbanks: University of Alaska Press.

———. 2003. *Sold American: The Story of Alaska Natives and their Land, 1867–1959.* Fairbanks: University of Alaska Press. First published 1997 by University Press of New England.

Mitchell, Ronald B. 1994. *Intentional Oil Pollution at Sea: Environmental Policy and Treaty Compliance.* Cambridge, MA: MIT Press.

Morton, Rogers C. B. 1972. *Statement by Secretary of the Interior Rogers C. B. Morton concerning Application for a Trans-Alaska Pipeline Right of Way.* May 11, 1972. Washington, DC: U.S. Department of the Interior.

Nash, Roderick Frazier. 2001. *Wilderness and the American Mind.* 4th ed. New Haven, CT: Yale University Press.

Naske, Claus-M., and Herman E. Slotnick. 2011. *Alaska: A History.* 3rd ed. Norman: University of Oklahoma Press.

National Institute of Standards and Technology. 2011. *Specifications, Tolerances, and Other Technical Requirements for Weighing and Measuring Devices as Adopted by the 96th National Conference on Weights and Measures 2011.* National Institute of Standards and Technology Handbook 44, 2012 ed. Washington, DC: U.S. Department of Commerce.

National Research Council. 1998. *Double-Hull Tanker Legislation: An Assessment of the Oil Pollution Act of 1990.* Washington, DC: National Academies Press.

———. 2002. *A Century of Ecosystem Science: Planning Long-Term Research in the Gulf of Alaska.* Washington, DC: National Academies Press.

———. 2003. *Cumulative Environmental Effects of Oil and Gas Activities on Alaska's North Slope.* Washington, DC: National Academies Press.

Nelson, Daniel. 2004. *Northern Landscapes: The Struggle for Wilderness Alaska.* Washington, DC: Resources for the Future Press.

———. 2009. *A Passion for the Land: John F. Seiberling and the Environmental Movement.* Kent, OH: The Kent State University Press.

Nixon, Richard. 1970. "Special Message to the Congress on Indian Affairs." July 8, 1970. In *Public Papers of the Presidents of the United States: Richard Nixon, 1970.* Washington, DC: U.S. Government Printing Office.

———. 1971a. "Special Message to the Congress Proposing the 1971 Environmental Program." February 8, 1971. In *Public Papers of the Presidents of the United States: Richard Nixon, 1971.* Washington, DC: U.S. Government Printing Office.

————. 1971b. "Statement about an Alaska Natives' Claims Bill." April 6, 1971. In *Public Papers of the Presidents of the United States: Richard Nixon, 1971.* Washington, DC: U.S. Government Printing Office.

————. 1973a. "Special Message to the Congress on Energy Policy." April 18, 1973. In *Public Papers of the Presidents of the United States: Richard Nixon, 1973.* Washington, DC: U.S. Government Printing Office.

————. 1973b. "Statement Announcing Additional Energy Policy Measures." June 29, 1973. In *Public Papers of the Presidents of the United States: Richard Nixon, 1973.* Washington, DC: U.S. Government Printing Office.

————. 1973c. "Special Message to the Congress Proposing Emergency Energy Legislation." November 8, 1973. In *Public Papers of the Presidents of the United States: Richard Nixon, 1973.* Washington, DC: U.S. Government Printing Office.

————. 1973d. "Statement about the Trans-Alaska Oil Pipeline." November 16, 1973. In *Public Papers of the Presidents of the United States: Richard Nixon, 1973.* Washington, DC: U.S. Government Printing Office.

NOAA (United States National Oceanic and Atmospheric Administration). 1998. *Proceedings of A Symposium on Practical Ice Observation in Cook Inlet and Prince William Sound.* Washington, DC: NOAA.

Norris, Frank. 2002. *Alaska Subsistence: A National Park Service Management History.* Anchorage: Alaska Support Office, U.S. National Park Service.

Norris, Frank, Julie Johnson, Sande Anderson, and Logan Hovis. 1999. *The Alaska Journey: One Hundred and Fifty Years of the Department of the Interior in Alaska.* Anchorage: Alaska Support Office, U.S. National Park Service.

NTSB (United States National Transportation Safety Board). 1990. *Marine Accident Report—Grounding of the U.S. Tankship Exxon Valdez on Bligh Reef, Prince William Sound Near Valdez, Alaska, March 24, 1989.* Washington, DC: U.S. Department of Commerce.

Oil Spill Commission Action. 2012. *Assessing Progress: Implementing the Recommendations of the National Oil Spill Commission.* Washington, DC: Oil Spill Commission Action.

OSRI (Prince William Sound Oil Spill Recovery Institute). 2003. *OSRI Two-year Report 2001–2002.* Cordova, AK: OSRI.

Parks Canada. 2007. *Ivvavik National Park of Canada Management Plan.* Quebec, Canada: Parks Canada.

————. 2010. *Kluane National Park and Reserve of Canada Management Plan.* Quebec, Canada: Parks Canada.

Patashnik, Eric M. 2008. *Reforms at Risk: What Happens After Major Policy Changes Are Enacted.* Princeton, NJ: Princeton University Press.

PWS RCAC (Prince William Sound Regional Citizens' Advisory Council). 1991a. *Annual Report.* Anchorage, AK: PWS RCAC.

———.1991b. *The Observer* 1(2). Anchorage, AK: PWS RCAC.

———. 1993a. *1993 A Year In Review.* Anchorage, AK: PWS RCAC.

———. 1993b. *The Observer* 3(2). Anchorage, AK: PWS RCAC.

———. 1993c. *The Observer* 3(3). Anchorage, AK: PWS RCAC.

———. 1993d. *Then and Now: Changes Since the Exxon Valdez Oil Spill.* Anchorage, AK: PWS RCAC.

———. 1994a. *The Observer* 4(1). Anchorage, AK: PWS RCAC.

———. 1994b. *The Observer* 4(2). Anchorage, AK: PWS RCAC.

———. 1994c. *The Observer* 4(4). Anchorage, AK: PWS RCAC.

———. 1995a. *The Observer* 5(2). Anchorage, AK: PWS RCAC.

———. 1995b. *The Observer* 5(4). Anchorage, AK: PWS RCAC.

———. 1995c. *Oil Spill Prevention: Improvements in Tanker Safety.* Anchorage, AK: PWS RCAC.

———. 1996a. *1996 A Year in Review.* Anchorage, AK: PWS RCAC.

———. 1996b. *The Observer* 6(3). Anchorage, AK: PWS RCAC.

———. 1997a. *The Observer* 7(1). Anchorage, AK: PWS RCAC.

———. 1997b. *The Observer* 7(2). Anchorage, AK: PWS RCAC.

———. 1997c. *The Observer* 7(3). Anchorage, AK: PWS RCAC.

———. 1998. *1997–1998 in Review.* Anchorage, AK: PWS RCAC.

———. 1999a. *The Observer* 9(2). Anchorage, AK: PWS RCAC.

———. 1999b. *1998–1999 In Review.* Anchorage, AK: PWS RCAC.

———. 2000a. *The Observer* 10(1). Anchorage, AK: PWS RCAC.

———. 2000b. *1999–2000 in Review.* Anchorage, AK: PWS RCAC.

———. 2001a. *2000–2001 in Review.* Anchorage, AK: PWS RCAC.

———. 2001b. *The Observer* 11(4). Anchorage, AK: PWS RCAC.

———. 2002a. *2001–2002 in Review.* Anchorage, AK: PWS RCAC.

———. 2002b. *The Observer* 12(1). Anchorage, AK: PWS RCAC.

———. 2002c. *The Observer* 12(2). Anchorage, AK: PWS RCAC.

———. 2003a. *The Observer* 13(1). Anchorage, AK: PWS RCAC.

———. 2003b. *2002–2003 Year in Review.* Anchorage, AK: PWS RCAC.

———. 2003c. *The Observer* 13(3). Anchorage, AK: PWS RCAC.

———. 2003d. *The Observer* 13(4). Anchorage, AK: PWS RCAC.

———. 2004a. *2003–2004 Year in Review.* Anchorage, AK: PWS RCAC.

———. 2004b. *The Observer* 14(3). Anchorage, AK: PWS RCAC.

———. 2005a. *The Observer* 15(1). Anchorage, AK: PWS RCAC.

———. 2005b. *The Observer* 15(3). Anchorage, AK: PWS RCAC.

———. 2005c. *2004–2005 in Review.* Anchorage, AK: PWS RCAC.

———. 2006. *The Observer* 16(2). Anchorage, AK: PWS RCAC.

———. 2007a. *2006–2007 in Review.* Anchorage, AK: PWS RCAC.

———. 2007b. *The Observer* 17(1). Anchorage, AK: PWS RCAC.

————. 2007c. *The Observer* 17(4). Anchorage, AK: PWS RCAC.

————. 2009a. *The Observer* 19(3). Anchorage, AK: PWS RCAC.

————. 2009b. *The Observer* 19(4). Anchorage, AK: PWS RCAC.

————. 2009c. *Then and Now: Changes in Prince William Sound Crude Oil Transportation Since the Exxon Valdez Oil Spill.* Anchorage, AK: PWS RCAC.

————. 2010. *The Observer* 20(1). Anchorage, AK: PWS RCAC.

————. 2011a. *The Observer* 21(4). Anchorage, AK: PWS RCAC.

————. 2011b. *2010–2011 Annual Report.* Anchorage, AK: PWS RCAC.

PWS Science Center (Prince William Sound Science Center). 2004. *Investing in Science for the Future of Prince William Sound: An Overview of Programs 1989–2003.* Cordova, AK: PWS Science Center.

Ramseur, Jonathan L. 2010. *Oil Spills in U.S. Coastal Waters: Background and Governance.* Washington, DC: Congressional Research Service.

Randle, Russell V. 2012. *Oil Pollution Deskbook.* 2nd ed. Washington, DC: Environmental Law Institute.

Reagan, Ronald. 1983a. "Statement on United States Oceans Policy." March 10, 1983. In *Public Papers of the Presidents of the United States: Ronald Reagan, 1983,* bk. 1. Washington, DC: U.S. Government Printing Office.

————. 1983b. "Proclamation 5030—Exclusive Economic Zone of the United States of America." March 10, 1983. In *Public Papers of the Presidents of the United States: Ronald Reagan, 1983,* bk. 1. Washington, DC: U.S. Government Printing Office.

Reiss, Bob. 2012. *The Eskimo and the Oil Man: The Battle at the Top of the World for America's Future.* New York: Business Plus.

Rennick, Penny, ed. 1993. "Prince William Sound." *Alaska Geographic* 20(1): 1–112.

Rennicke, Jeff. 1995. "Heritage of the North." In *Exploring Canada's Spectacular National Parks* by David Dunbar, Tom Melham, Lawrence Millman, Cynthia Russ Ramsay, Jeff Rennicke, and Jennifer C. Urquhart. Washington, DC: National Geographic Society.

Repetto, Robert, ed. 2006. *Punctuated Equilibrium and the Dynamics of U.S. Environmental Policy.* New Haven, CT: Yale University Press.

Ross, Ken. 2000. *Environmental Conflict in Alaska.* Boulder: University Press of Colorado.

Rothman, Hal. 1994. *America's National Monuments: The Politics of Preservation.* Lawrence: University Press of Kansas. First published 1989 by University of Illinois Press.

Runte, Alfred. 2010. *National Parks: The American Experience.* 4th ed. Lanham, MD: Taylor Trade Publishing.

Shah, Monica, ed. 2005. "The Legacy of ANILCA." *Alaska Park Science* 4(2): 1–51.

Siehl, George H., and Robert S. Want. 1972. *Highlights of the Department of the Interior's Environmental Report and Decision on the Proposed Trans-Alaska Oil Pipeline.* Washington, DC: Congressional Research Service.

Skillen, James R. 2009. *The Nation's Largest Landlord: The Bureau of Land Management in the American West.* Lawrence: University Press of Kansas.

Smith, Thomas G. 2006. *Green Republican: John Saylor and the Preservation of America's Wilderness.* Pittsburgh: University of Pittsburgh Press.

Sohn, Louis B., Kristen Gustafson Juras, John E. Noyes, and Erik Franckx. 2010. *Law of the Sea in a Nutshell.* 2nd ed. St. Paul, MN: Thomson/West.

Spence, Mark David. 1999. *Dispossessing the Wilderness: Indian Removal and the Making of the National Parks.* New York: Oxford University Press.

S. Rep. (U.S. Senate Report) No. 93-207. 1973. *Federal Lands Right-of-Way Act of 1973.* Washington, DC: U.S. Government Printing Office.

S. Rep. (U.S. Senate Report) No. 94-593. 1976. *The National Wildlife Refuge System Administration Act Amendments.* Washington, DC: U.S. Government Printing Office.

S. Rep. (U.S. Senate Report) No. 95-1300. 1978. *Designating Certain Lands in the State of Alaska as Units of the National Park, National Wildlife Refuge, National Wild and Scenic Rivers, and National Wilderness Preservation Systems, and for Other Purposes.* Washington, DC: U.S. Government Printing Office.

S. Rep. (U.S. Senate Report) No. 96-413. 1979. *Alaska National Interest Lands.* Washington, DC: U.S. Government Printing Office.

S. Rep. (U.S. Senate Report) No. 101-94. 1989. *Oil Pollution Liability and Compensation Act of 1989.* Washington, DC: U.S. Government Printing Office.

Stevens, Stan, ed. 1997. *Conservation through Cultural Survival: Indigenous Peoples and Protected Areas.* Washington, DC: Island Press.

Sutter, Paul S. 2002. *Driven Wild: How the Fight against Automobiles Launched the Modern Wilderness Movement.* Seattle: University of Washington Press.

Swedlow, Brendon. 2011. "Cultural Surprises as Sources of Sudden, Big Policy Change." *PS: Political Science & Politics* 44(4): 736–39.

Swem, Theodore R., and Robert Cahn. 1984. "The Politics of Parks in Alaska: Innovative Planning and Management Approaches for New Protected Areas." In *National Parks, Conservation, and Development: The Role of Protected Areas in Sustaining Society,* edited by Jeffrey A. McNeely and Kenton R. Miller, 518–24. Washington, DC: Smithsonian Institution Press.

Tan, Alan Khee-Jin. 2006. *Vessel-Source Marine Pollution: The Law and Politics of International Regulation.* Cambridge, UK: Cambridge University Press.

Tatshenshini-Alsek Management Board. 2001. *Management Direction Statement for Tatshenshini-Alsek Park.* Victoria, Canada: BC Parks.

Theberge, John B. 1978. "Kluane National Park." In *Northern Transitions Vol. I: Northern Resource and Land Use Policy Study*, edited by Everett B. Peterson and Janet B. Wright, 153–89. Ottawa, Canada: Canadian Arctic Resources Committee.

The Wilderness Society. 2001. *Alaska National Interest Lands Conservation Act Citizens' Guide*. Washington, DC: The Wilderness Society.

The Wilderness Society, Environmental Defense Fund, and Friends of the Earth, eds. 1972. *Comments on the Environmental Impact Statement for the Trans-Alaska Pipeline*. 4 vols. Washington, DC: Center for Law and Social Policy.

True, James L., Bryan D. Jones, and Frank R. Baumgartner. 2007. "Punctuated-Equilibrium Theory: Explaining Stability and Change in Public Policymaking." In *Theories of the Policy Process*, 2nd ed., edited by Paul A. Sabatier, 155–87. Boulder, CO: Westview Press.

Truman, Harry S. 1945a. *Proclamation 2667—Policy of the United States with Respect to the Natural Resources of the Subsoil and Sea Bed of the Continental Shelf*. 10 Fed. Reg. 12303 (September 28, 1945).

———. 1945b. *Proclamation 2668—Policy of the United States with Respect to Coastal Fisheries in Certain Areas of the High Seas*. 10 Fed. Reg. 12304 (September 28, 1945).

Turner, James Morton. 2007. "The Politics of Modern Wilderness." In *American Wilderness: A New History*, edited by Michael Lewis, 243–61. New York: Oxford University Press.

Unocal. 1987. *Oil Spill Contingency Plan for Cook Inlet, Alaska*. Anchorage, AK: Unocal Corporation.

U.S. Congressional Record. 1971. *United States of America Congressional Record: Proceedings and Debates of the 92d Congress First Session*. Washington, DC: U.S. Government Printing Office.

———. 1973. *United States of America Congressional Record: Proceedings and Debates of the 93d Congress First Session*. Washington, DC: U.S. Government Printing Office.

———. 1978. *United States of America Congressional Record: Proceedings and Debates of the 95th Congress Second Session*. Washington, DC: U.S. Government Printing Office.

———. 1979. *United States of America Congressional Record: Proceedings and Debates of the 96th Congress First Session*. Washington, DC: U.S. Government Printing Office.

———. 1980. *United States of America Congressional Record: Proceedings and Debates of the 96th Congress First Session*. Washington, DC: U.S. Government Printing Office.

———. 1989. *United States of America Congressional Record: Proceedings and Debates of the 101st Congress First Session*. Daily ed. Library of Congress THOMAS website. Accessed October 14, 2011. http://thomas.loc.gov/home/thomas.php.

———. 1990. *United States of America Congressional Record: Proceedings and Debates of the 101st Congress Second Session.* Daily ed. Library of Congress THOMAS website. Accessed October 14, 2011. http://thomas.loc.gov/home/thomas.php.

U.S. Department of the Interior. 1971. *Draft Environmental Impact Statement for the Trans-Alaska Pipeline.* Washington, DC: U.S. Department of the Interior.

———. 1972. *Final Environmental Impact Statement Proposed Trans-Alaska Pipeline.* 6 vols. Washington, DC: U.S. Department of the Interior.

———. 1987. *Arctic National Wildlife Refuge, Alaska, Coastal Plain Resource Assessment: Report and Recommendation to the Congress of the United States and Final Legislative Environmental Impact Statement.* Washington, DC: U.S. Department of the Interior.

———. 2008. *Secretary Kempthorne Announces Decision to Protect Polar Bears under Endangered Species Act.* May 14, 2008. Washington, DC: U.S. Department of the Interior.

U.S. Fish and Wildlife Service. 1964. *A Report on Fish and Wildlife Resources Affected by Rampart Canyon Dam and Reservoir Project, Yukon River, Alaska.* Juneau, AK: U.S. Fish and Wildlife Service.

———. 1988. *Arctic National Wildlife Refuge Final Comprehensive Conservation Plan, Environmental Impact Statement, Wilderness Review, and Wild River Plans.* Anchorage, AK: U.S. Fish and Wildlife Service.

———. 2008a. *Endangered and Threatened Wildlife and Plants; Determination of Threatened Status for the Polar Bear* (Ursus maritimus) *throughout Its Range.* 73 Fed. Reg. 28212 (May 15, 2008).

———. 2008b. "Time Line: Establishment and Management of the Arctic Refuge." Accessed June 28, 2012. http://arctic.fws.gov/timeline.htm.

———. 2010. *Endangered and Threatened Wildlife and Plants; Designation of Critical Habitat for the Polar Bear* (Ursus maritimus) *in the United States.* 75 Fed. Reg. 76086 (December 7, 2010).

Vale, Thomas R. 2005. *The American Wilderness: Reflections on Nature Protection in the United States.* Charlottesville: University of Virginia Press.

Wayburn, Edgar, and Allison Alsup. 2004. *Your Land and Mine: Evolution of a Conservationist.* San Francisco, CA: Sierra Club Books.

Whitehead, John S. 2004. *Completing the Union: Alaska, Hawai'i, and the Battle for Statehood.* Albuquerque: University of New Mexico Press.

Wilder, Robert Jay. 1998. *Listening to the Sea: The Politics of Improving Environmental Protection.* Pittsburgh: University of Pittsburgh Press.

Wilkinson, Cynthia M., L. Pittman, and Rebecca F. Dye. 1992. "Slick Work: An Analysis of the Oil Pollution Act of 1990." *Journal of Energy, Natural Resources, and Environmental Law* 12: 181–236.

Williams, Deborah. 1997. "ANILCA: A Different Legal Framework for Managing the Extraordinary National Park Units of the Last Frontier." *Denver University Law Review* 74(3): 859–68.

Willis, Roxanne. 2010. *Alaska's Place in the West: From the Last Frontier to the Last Great Wilderness.* Lawrence: University Press of Kansas.

Williss, G. Frank. 2005. *"Do Things Right the First Time": The National Park Service and the Alaska National Interest Lands Conservation Act of 1980.* 2nd ed. Anchorage: Alaska Regional Office, U.S. National Park Service.

Woods, Bruce. 2003. "Alaska's National Wildlife Refuges." *Alaska Geographic* 30(1): 1–96.

Worster, Donald. 1994. *Under Western Skies: Nature and History in the American West.* New York: Oxford University Press. First published 1992 by Oxford University Press.

INDEX

Note: Figures are represented by boldface page numbers followed by "f" (e.g., **29f**). Maps are represented by boldface page numbers in parentheses **(35)**.

157